ELECTRONIC PRINCIPLES
AND APPLICATIONS

Revision and Self Assessment Series

ELECTRONIC PRINCIPLES AND APPLICATIONS

JOHN B. PRATLEY

BA MEd CEng MIEE
Academic Tutor and Consultant
Bridgnorth, Shropshire

City of Westminster College
Paddington Learning Centre
25 Paddington Green
London W2 1NB

A member of the Hodder Headline Group
LONDON • SYDNEY • AUCKLAND

First published in Great Britain in 1998 by Arnold,
a member of the Hodder Headline Group,
338 Euston Road, London NW1 3BH.

http://www.arnoldpublishers.com

British Library Cataloguing in Publication Data
A catalogue record for this book is available from the British Library.

ISBN 0 340 69275 8

Publisher: Dilys Alam
Production Editor: Wendy Rooke
Production Controller: Rose James
Cover Designer: Terry Griffiths

Typeset in 10/12 Adobe Garamond by Alden Bookset, Didcot
Printed and bound in Great Britain by J. W. Arrowsmith Ltd., Bristol

Dedicated to my wife and family who have taught me a lot.

CONTENTS

PREFACE

The purpose of this book is to teach electronics to a beginner, giving an intensive introduction to the electrical and electronic principles and practices of everyday modern electronics. No previous knowledge or experience of electronics is assumed, but it is quite helpful if you have some numerical ability.

The contents of the book have been chosen so that they cover the requirements of the GNVQ Intermediate Engineering course in Electronics. It will also be useful for GCSE Electronics, parts of Advanced Level Electronics, Certificate and Diploma courses in Electrical and Electronic Engineering and parts of CGLI Electronics courses.

Learning does not take place in Electrical and Electronic Principles by just reading about it. Reading is the starting point, but then you need to become active. Each of the initial chapters includes worked examples. These have been included to help with the understanding of the preceding theory. Following the worked examples you will find an abundance of exercises for you to attempt. If you get stuck, remember there is always a teacher to help you. Further help is available because answers are provided to most of the questions.

A vitally important section of the book comes in the later chapters where more than 50 investigations are provided. The investigations provide the student with hands-on experience of the components dealt with in previous chapters. In all cases a general list of components and equipment is provided. Specific details are left to the resources of the teacher and student.

Circuit diagrams are provided and it is assumed that in most cases a breadboard will be used. However, if breadboards are not available the investigations can still be carried out using other forms of board. Remember there are many sources of information available to you, not least through manufacturers' catalogues and data sheets. After you have completed an investigation you will find at the end of the chapter a summary of what you should have found.

I do hope that you do not find any mistakes in the book but if you do please inform the publisher who will pass them on to me. On this matter some comfort is taken in remembering the great author Daniel Defoe when he wrote *Robinson Crusoe*. Defoe stated that 'our hero stripped NAKED, swam out to the ship that he had recently escaped from, filled his POCKETS with items that might help him to stay alive on the desert island'. Even the great make some mistakes so there might be hope for me yet.

Extracts from British Standards are reproduced with the permission of BSI. Complete editions of the standards can be obtained by post from BSI Customer Services, 389 Chiswick High Road, London W4 4AL.

Finally, I hope the book gives pleasure and helps you to learn.

John B. Pratley

TABLE OF TERMS

Quantity	Symbol	Unit	Unit abbreviation
Angular velocity	ω	radian per second	rad/s
Area	a	square metre	m^2
Capacitance	C	farad	F
Charge	Q	coulomb	C
Current			
steady or r.m.s. value	I	ampere	A
instantaneous value	i		
maximum value	I_m		
Current gain of amplification factor			
common base circuit	h_{fb}		
common emitter circuit	h_{fe}		
Distance	d	metre	m
Electric current			
base	I_b	ampere	A
		milliampere	mA
		microampere	μA
collector	I_c	ampere	A
		milliampere	mA
		microampere	μA
emitter	I_e	ampere	A
		milliampere	mA
		microampere	μA
Electric field strength	E	volt/metre	V/m
Electric flux density	D	coulomb/square metre	C/m^2
Electromative force	V	volt	V
Energy	W	joule	J
Force	F	newton	N
Frequency	f	hertz	Hz
Inductance	L	henry	H
Length	l	metre	m
Magnetic flux	Φ	weber	Wb
Magnetic flux density	B	tesla	T
Magnetizing force	H	ampere/metre	A/m
Number	n		
Periodic time	T	second	s
Permittivity			
absolute	ϵ	farad/metre	F/m
of free space	ϵ_0	farad/metre	F/m
relative	ϵ_r		

Phase angle	ϕ	degree	°
Plane angle	θ	degree	°
Potential difference			
steady or r.m.s. value	V	volt	V
instantaneous value	v		
maximum value	V_m		
Radius	r	metre	m
Resistance	R	ohm	Ω
Resistivity	ρ	ohm metre	Ωm
Temperature*	θ	kelvin	K
Time	t	second	s
Time constant	T	second	s
Velocity	v	metre/second	m/s
Voltage			
base-emitter	V_{be}	volt	V
		millivolt	mV
		microvolt	μV
collector-base	V_{cb}	volt	V
		millivolt	mV
		microvolt	μV
collector-emitter	V_{ce}	volt	V
		millivolt	mV
		microvolt	μV

---

* Temperature difference is commonly expressed in degrees Celsius (°C) instead of kelvins. It is useful that the unit of Celsius and kelvin scales is the same, i.e.

$$1 \text{ degree Celsius} = 1 \text{ kelvin}$$

A temperature expressed in degrees Celsius is equal to the temperature expressed in kelvins less 273, i.e.

$$400 \text{ K} = 400 - 273 = 127°\text{C}$$

1

PRINCIPLES OF ELECTRICITY

1.1 ELECTRIC CURRENT

An electric current (I) measured in amperes (A) is the result of the movement of electrons through a material in one direction.

1.2 ELECTROMOTIVE FORCE

For electric current to flow in a circuit, two conditions are necessary: a continuous circuit and a force to move the electrons. A device such as a simple cell provides the force to move the electrons round the circuit. The maximum force available is given a special name: electromotive force (abbreviation e.m.f.) (E), measured in volts (V).

1.3 RESISTANCE

Loads such as lamps, water heaters, etc. when placed in an electrical circuit oppose the flow of electrons. This opposition is called resistance (R), and is measured in ohms (Ω).

1.4 OHM'S LAW

Resistance is equal to the voltage at the terminals of a resistor divided by the current flowing through it, i.e.

$$\text{resistance} = \frac{\text{voltage}}{\text{current}} \text{ or } R = \frac{V}{I}$$

The voltage is given the name potential difference.

It is measured in terms of the work done in transferring a charge from one point to another.

Ohm's law states that the potential difference across the terminals of a pure resistor is directly proportional to the current flowing through the resistor. That is, potential difference (V) $\propto$ current (I) and

$$V = IR$$

Examples

1. Calculate the resistance of a coil with a potential difference across it of 12 V and a current flow of 2 A.

$$V = 12\,\text{V} \quad I = 2\,\text{A}$$

Using $R = V/I$

$$R = \frac{12}{2} = 6\,\Omega$$

2. A potential difference of 240 V is applied across the terminals of a 30 Ω resistor. Determine the current flowing through the resistor.

$$V = 240\,\text{V} \quad R = 30\,\Omega$$

Using $I = V/R$

$$I = \frac{240}{30} = 8\,\text{A}$$

3. Calculate the potential difference across a lamp if its resistance is 50 Ω and the current flowing through it is 2 A.

$$R = 50\,\Omega \quad I = 2\,\text{A}$$

Using $V = IR$

$$V = 2 \times 50 = 100\,\text{V}$$

Exercise 1.1

1. An electric heating element has a resistance of 40 Ω and is to operate on a 240 V direct current (d.c.) supply. Calculate the current taken by the heater.

2. A cable of resistance of 0.04 Ω carries a current of 50 A. Determine the potential difference across the ends of the cable.

3. A lamp is rated at 12 V 3.5 A. What is its resistance?

4. What will be the current flow when a 24 Ω resistor is connected across a 2 V supply?

5. A solenoid has a resistance of 400 Ω. Calculate the current through the solenoid when it is connected across a 120 V supply.

6. A 12 V battery is connected across a 3.5 Ω resistor. Calculate the current flowing.

7. A lamp is connected to a 240 V supply. The current taken when it is first switched on is 2 A, and when the filament becomes hot is 1.5 A. Determine the resistance of the lamp initially and when it is hot.

8. The current through the field windings of a motor is 1.8 A when the resistance is 100 Ω. After five minutes the resistance rises to 120 Ω. If the potential difference remains constant determine the new value of current flow.

9. A contactor coil of resistance 200 Ω has a current flowing through it of 0.5 A. If the supply voltage remains constant determine the current flow when the coil resistance is: (a) 210 Ω; (b) 220 Ω; (c) 230 Ω; (d) 240 Ω; (e) 250 Ω.

10. The current flowing through a resistor is 5 A when the potential difference across it is 60 V. Determine: (a) the potential difference when the current is 7 A; (b) the current when the potential difference is 35 V.

11. A length of two core cable has a resistance of 0.4 Ω for each conductor. What will be the voltage drop when the cable carries a current of 25 A?

12. A circuit has a resistance of 20 Ω. Calculate: (a) the current flowing when a potential difference of 260 V is applied to the circuit; (b) the current which would flow if the applied potential difference were reduced to 200 V. What added resistance is required to reduce the current to 5 A when 260 V are applied?

13. If a voltmeter has a resistance of 25 kΩ, calculate the current absorbed when it is connected across a 415 V supply.

1.5 LINEAR AND NON-LINEAR COMPONENTS

A graph of the relationship between potential difference (V) and current (I) for a single linear resistor may be obtained by investigation using the circuit in Fig. 1.1.

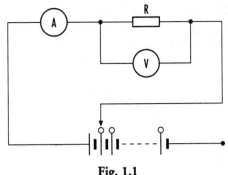

Fig. 1.1

It is important to note that the ammeter is always connected in series with the component and that the voltmeter is connected across the component. The potential difference across the resistor is varied by using a d.c. variable supply. A typical set of results is shown in Table 1.1.

Table 1.1

Potential difference (V)	0	2	4	6	8	10	12
Current (A)	0	0.1	0.2	0.3	0.4	0.5	0.6

The graph of potential difference (V) versus current (I) is shown in Fig. 1.2.

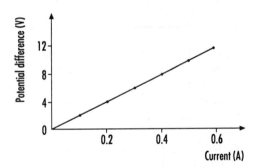

Fig. 1.2

A graph of the relationship between potential difference (V) and current (I) for a non-linear component, such as a lamp, may be obtained by investigation using the circuit as in Fig. 1.1 except that the lamp replaces the resistor. A typical set of results is shown in Table 1.2.

Table 1.2

Potential difference (V)	0	2	4	6	8	10	12
Current (A)		0	0.02	0.04	0.11	0.22	0.38	0.60

The graph of potential difference versus current is shown in Fig. 1.3.

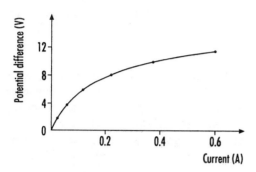

Fig. 1.3

Exercise 1.2_____

1. A test on a resistor gave the following results:

Potential difference (V)	0	25	50	75	100	125	150	175
Current (A)		0	1	2	3	4	5	6	7

Plot the graph showing the relationship between the potential difference and the current.

Is the resistor used in the test linear or non-linear? Give a reason for your answer.

2. The following readings of potential difference and current relate to an electric heater element:

Potential difference (V)	0	5	10	15	20	25	30	35
Current (A)		0	1.5	3.0	4.5	6.0	7.5	9.0	10.5

Plot a graph and determine from it: (a) the resistance of the electric heater element; (b) the current when the potential difference is 17 V; (c) the potential difference when the current is 9.5 A.

3. A test on a lamp gave the following results:

Potential difference (V)	0	3	6	9	12	15	18
Current (A)		0	0.1	0.4	1.5	3.3	5.6	8.0

Plot a graph and from it deduce whether the resistance of the lamp is linear or non-linear. Give a reason for your choice.

4. A test on a linear and a non-linear resistor gave the following readings:

Potential difference (V)	0	1	2	3	4	5	6
Linear resistance Current (mA)	0	4.0	8.0	12.0	16.0	20.0	24.0
Non-linear resistance Current (mA)	0	1.0	2.0	12.5	20.5	35.5	52.0

Plot graphs of current to a common base of potential difference.

Comment on the shapes of the graphs obtained.

For each value of potential difference calculate the resistance of the linear and non-linear resistor and then plot graphs of resistance to a common base of potential difference.

Comment on the shapes of the graphs obtained.

5. The following table shows the current and voltage readings obtained in an experiment to find the value of an unknown constant resistance. Draw a circuit diagram to show the apparatus that would be used for this experiment.

Plot the results on graph paper, and from the graph find the value of the resistance.

Find how much power this resistor will dissipate when connected to a battery of voltage 4 V.

V (V)	2.2	5	7	9.4	11
I (A)	1	2	3	4	4.5

1.6 ELECTRICAL POWER

Power (P) is defined as the rate of consuming or producing energy, i.e. energy used (W)/time (t) or

$$P = \frac{W}{t}$$

Therefore

$$P = \frac{VIt}{t} \quad P = VI$$

or

$$P = \frac{I^2 Rt}{t} \quad P = I^2 R$$

or

$$P = \frac{V^2 t}{Rt} \quad P = \frac{V^2}{R}$$

Power is measured in watts (W), kilowatts (kW), or megawatts (MW).

Examples

1. Calculate the power input to a motor taking 8 A from a 250 V d.c. supply.

$$I = 8\,\text{A} \quad V = 250\,\text{V}$$

Using $P = VI$

$$P = 250 \times 8 = 2000\,\text{W} = 2\,\text{kW}$$

2. A cable carrying a current of 20 A has a total resistance of 1.5 Ω. Determine the power absorbed by the cable.

$$I = 20\,\text{A} \quad R = 1.5\,\Omega$$

Using $P = I^2 R$

$$P = 20 \times 20 \times 1.5 = 600\,\text{W}$$

3. Determine the power absorbed by a 20 Ω resistor connected across a 2 V d.c. supply.

$$R = 20\,\Omega \quad V = 2\,\text{V}$$

Using $P = V^2/R$

$$P = \frac{2 \times 2}{20} = \frac{4}{20} = 0.2\,\text{W} = 200\,\text{mW}$$

4. Calculate the current flowing through a 4 Ω resistor which dissipates 100 W.

$$R = 4\,\Omega \quad P = 100\,\text{W}$$

From $P = I^2 R$, $\quad I^2 = P/R \quad$ and $I = \sqrt{(P/R)}$

$$I = \sqrt{\left(\frac{100}{4}\right)} = \sqrt{25} = 5\,\text{A}$$

Exercise 1.3

1. An electric iron element carries a current of 2.5 A when connected across a 240 V supply. Calculate the power used.

2. Calculate the current taken from a 2 kW heater being used on a 250 V supply.

3. Calculate the potential difference across a resistor carrying a current of 27 A and consuming 8 kW.

4. Calculate the power absorbed by a heater of resistance 10 Ω when a current of 5 A is flowing.

5. Calculate the current flowing through a 500 Ω resistor rated at 8 kW.

6. A current of 1.6 A is flowing through a resistor. Determine the value of the resistor when the power absorbed is 60 W.

7. A coil has a resistance of 700 Ω. Calculate the power absorbed by this coil from a 24 V d.c. supply.

8. Calculate the resistance of a 48 W 12 V lamp.

9. Calculate the potential difference across a 65 Ω resistor rated at 20 W.

10. Complete the table below. All of the appliances are connected to a 240 V supply.

Appliance	Power rating	Current (A)
Fire	2 kW	
Iron	200 W	
Resistor	5 W	
Toaster	550 W	
Coffee percolator	250 W	
Immersion heater	3 kW	
Kettle	2.5 kW	
Lamp	100 W	
Radiator	6 kW	
Storage heater	8 kW	

1.7 CIRCUIT DIAGRAMS

A circuit diagram is a diagram which shows by means of symbols the components and their interconnections concerned in the operation of a circuit. The aim should be to show the operation of the circuit as clearly as possible; circuit diagrams do not necessarily show the best physical layout of the components and their connections. A simple series circuit is shown in Fig. 1.4.

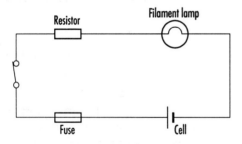

Fig. 1.4

When making circuit diagrams in electrical engineering it is necessary to use symbols to denote the various components. We use the internationally agreed British Standard graphical symbols, which have the advantage that anyone examining a diagram can recognize the symbols used, thus making it unnecessary to refer to a new list of symbols on each occasion.

A selection of British Standard graphical symbols is shown in Fig. 1.5. A complete list of graphical symbols for electrical and electronic diagrams can be found in British Standards.

Description	Symbol
Primary or secondary cell	
Battery of primary or secondary cells: alternative symbol	
Battery with tappings	
Positive polarity	
Negative polarity	
Fixed resistor: general symbol	
Fixed resistor with fixed tappings	
Variable resistor: general symbol	
Resistor with preset adjustment	
Resistor with moving contact	
Heater	
Signal lamp: general symbol	
Filament lamp	
Discharge lamp: general symbol	
Cold cathode discharge lamp, e.g. neon lamp	
Hot cathode tubular fluorescent lamp	
Earth	
Electric bell: general symbol	
Electric buzzer	
Siren	

Fig. 1.5

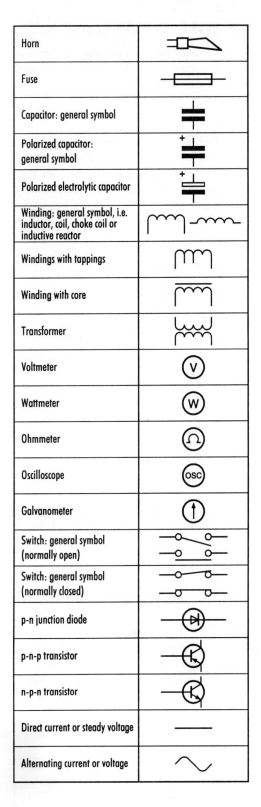

Horn	
Fuse	
Capacitor: general symbol	
Polarized capacitor: general symbol	
Polarized electrolytic capacitor	
Winding: general symbol, i.e. inductor, coil, choke coil or inductive reactor	
Windings with tappings	
Winding with core	
Transformer	
Voltmeter	
Wattmeter	
Ohmmeter	
Oscilloscope	
Galvanometer	
Switch: general symbol (normally open)	
Switch: general symbol (normally closed)	
p-n junction diode	
p-n-p transistor	
n-p-n transistor	
Direct current or steady voltage	
Alternating current or voltage	

Fig. 1.5 *(cont.)*

The simple series circuit in Fig. 1.4 shows the lines as conductors with five components: a resistor, a cell, a fuse, a switch and a filament lamp. In more complicated circuits, lines representing conductors may have to branch out or cross. A British Standard convention is in use to avoid misinterpreting the meaning of a circuit diagram. The methods used for showing how conductors cross and meet at joints are shown in Fig. 1.6.

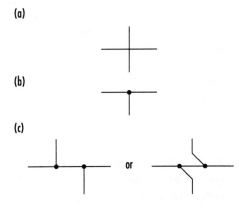

Fig. 1.6 (a) Conductors crossing but not joined, (b) junction of conductors, (c) double junction of conductors.

Exercise 1.4

1. Draw the following British Standard graphical symbols: cell, filament lamp, fuse and resistor.

2. Show two methods of making a joint on a circuit diagram.

3. Show how crossings are made in conductors on a circuit diagram.

4. Sketch five types of resistor symbol for use in an electrical circuit giving an application for each type.

5. Draw the general symbol for a switch in the open and closed position.

6. Complete the series circuit diagram so that the potential difference across the lamp can be measured.

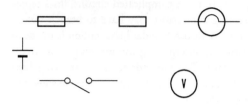

7. Draw a circuit diagram, using BS graphical symbols, showing a bell, fuse, cell and a switch in the open position connected in series.

8. Explain the fundamental difference between the two resistors shown.

9. The sketch shows an outline circuit diagram. Redraw the circuit and label the component parts as shown by the arrows.

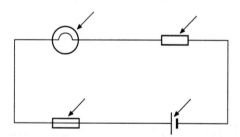

10. Complete the circuit diagram so that the three resistors are connected in series and the total current flow can be measured by an ammeter and the potential difference across the cell can be measured by a voltmeter.

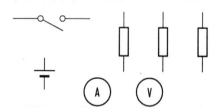

11. Draw a circuit diagram showing a cell, a fuse, a bell push and a bell connected in series. Include in the circuit an ammeter to measure total current flow and a voltmeter to measure the potential difference of the cell.

1.8 WIRING DIAGRAMS

A wiring diagram shows the connections between components and indicates the physical layout of the components.

Example

1. Examine the given circuit diagram. Complete the wiring diagram in the plan view provided. The conductors must not cross one another.

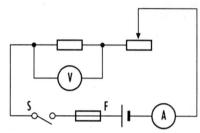

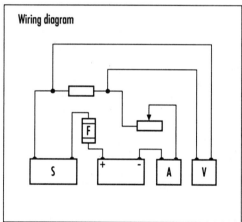

Wiring diagram

Exercise 1.5

1. Examine Fig. 1.4 then complete the wiring diagram in the plan view of the container with the layout given. Draw the container to a size of 12 cm × 6 cm. The conductors must not cross.

Wiring diagram layout

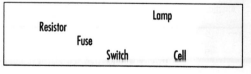

2. Examine the figure in Question 10 on p. 8. Redraw the plan view of the container to a size of 10 cm × 6 cm and wire up the circuit without any conductors crossing.

Wiring diagram layout

Resistor	Resistor		
	Resistor		Voltmeter
	Ammeter	Switch	Cell

3. The figure shows the circuit diagram for a single stage audio-frequency amplifier. Construct a wiring diagram for the given layout without any conductors crossing.

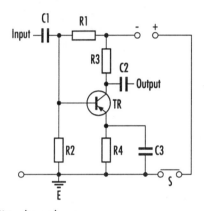

Wiring diagram layout

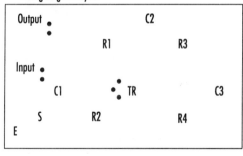

4. The figure shows a circuit diagram. Draw a wiring diagram with the minimum number of conductors crossing.

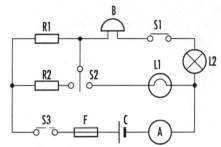

Wiring diagram layout

	R1	R2		S3
	F		B	S2
			L1	•••
+	−			S1
C	A		L2	

Answers

Exercise 1.1

1. 6 A
2. 2 V
3. 3.43 Ω
4. 0.083 A
5. 0.3 A
6. 3.429 A
7. 120 Ω
   160 Ω
8. 1.5 A
9. 0.48 A
   0.46 A
   0.44 A
   0.42 A
   0.4 A
10. 84 V
    2.92 A
11. 20 V
12. 13 A
    10 A
    32 Ω
13. 16.6 mA

Exercise 1.2

1. Linear – V directly proportional to I
2. $3.33\,\Omega$
   5.1 A
   31.67 V
3. Non-linear – V is not proportional to I
5. $2.4\,\Omega$
   6.66 W

Exercise 1.3

1. 600 W
2. 8 A
3. 296.3 V
4. 250 W
5. 4 A

6. $23.44\,\Omega$
7. 0.82 W
8. $3\,\Omega$
9. 36.1 V
10. 8.33 A
    0.83 A
    0.021 A
    2.29 A
    1.042 A
    12.5 A
    10.42 A
    0.42 A
    25 A
    33.33 A

2

ALTERNATING VOLTAGE AND CURRENT

2.1 DEGREES AND RADIANS

An angle is the amount of rotation between two straight lines and is measured in degrees, there being 360 degrees to a complete revolution.

A radian is defined as the angle subtended at the centre of a circle by an arc equal in length to the radius of the circle, as shown in Fig. 2.1.

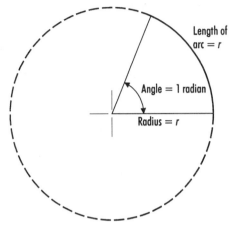

Fig. 2.1.

The circumference of a circle is given by $2\pi r$, and its entirety subtends an angle of $360°$ at the centre. Therefore

$$2\pi \text{ radians} = 360°$$

and

$$1 \text{ radian} = \frac{360°}{2\pi} = 57.3°$$

Examples

1. Convert $\pi/3$ radians to degrees.

$$\pi = 180°$$

Then

$$\frac{\pi}{3} = \frac{180}{3} = 60°$$

2. Convert $105°$ to radians.

$$180° = \pi \text{ radians} \quad 1° = \frac{\pi}{180} \text{ radians}$$

Then

$$105° = \frac{\pi}{180} \times 105 = 1.8326 \text{ radians}$$

3. Convert 3.864 radians to degrees.

$$1 \text{ radian} = 57.3°$$

Therefore

$$3.864 \text{ radians} = 3.864 \times 57.3$$

$$= 221.4072°$$

or $221.4°$ correct to 1 decimal place.

2.2 GENERATED E.M.F.

A simple alternating current (a.c.) generator consists of a loop coil and a permanent magnet as shown in Fig. 2.2. The ends of the coil are connected to two slip rings. Pressing against the slip rings are carbon brushes. When the coil rotates in the permanent magnet it cuts the magnet's field and an e.m.f. is induced in the coil.

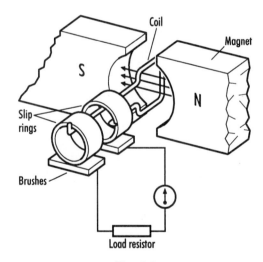

Fig. 2.2

Consider a single coil making one revolution or one cycle in an anticlockwise direction as shown in Fig. 2.3(a). The shape of the waveform generated is shown in Fig. 2.3(b).

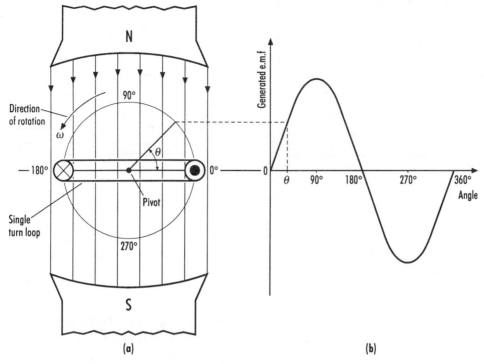

Fig. 2.3

This generated e.m.f. varies sinusoidally with time. It is more simply called a sine wave. The half-wave above the axes is said to be positive and the half-wave below the axes negative. The first positive peak value occurs 90° after the start of the cycle, and the first negative peak value at 270°. The e.m.f. generated in the coil causes a current to flow through a circuit connected by means of the slip rings and the brushes. A current will flow only if there is a load. In this case the resistor is the load and a current will flow. The current waveform is the same shape as that of the e.m.f.

In practice, a real generator has several coils wound in evenly-spaced slots in a soft iron cylinder and the permanent magnet is replaced by an electromagnet.

2.3 MAXIMUM AND INSTANTANEOUS VALUES

In Fig. 2.4 the maximum value (V) and the instantaneous e.m.f. (v) at any instant in time is shown. This is also true for maximum current (I) and instantaneous values of current (i). There is a relationship between the maximum

value, the instantaneous value and the angle $\theta°$ that the coil has turned through. It is $v = V \text{ sine } \theta$ usually written as $v = V \sin \theta$. (If you look on your calculator you should find the expression sin. More about this later.)

Likewise, the instantaneous current, i, is given by $i = I \sin \theta$.

For the record, sometimes the maximum value is also called the peak value or the amplitude.

Examples

1. An alternating e.m.f. is represented by $v = 100 \sin \theta$. Determine the value of the instantaneous voltage v when the angle θ is 30°.

$$\theta = 30°$$

Using $v = 100 \sin \theta$

$$v = 100 \sin 30°$$

From the calculator $\sin 30° = 0.5$ and so

$$v = 100 \times 0.5 = 50 \,\text{V}$$

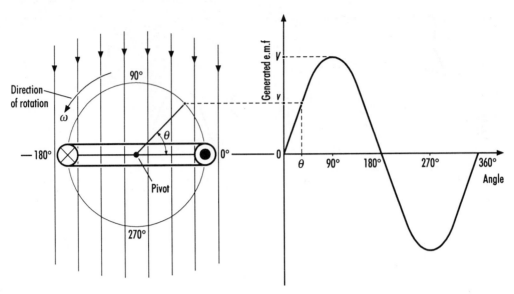

Fig. 2.4

2. An a.c. is represented by $i = 25 \sin \theta$. Determine the value of $\sin \theta$ and θ if the instantaneous value of current is 5 A.

$$I = 5\,\text{A}$$

Using $i = 25 \sin \theta$

$$5 = 25 \sin \theta$$

So

$$\frac{5}{25} = \sin \theta$$

$$\sin \theta = 0.2$$

and

$$\theta° = \text{inverse } \sin 0.2 = 11.54$$

3. The current produced by an alternating e.m.f. has an amplitude of 40 A. Determine the instantaneous value of the current when the loop is 45° anticlockwise from the horizontal.

$$I = 40\,\text{A} \quad \theta = 45°$$

Using $i = I \sin \theta$

$$i = 40 \sin 45° = 40 \times 0.7071$$

$$- 28.284\,\text{A}$$

2.4 PERIODIC TIME AND FREQUENCY

The periodic time (T) is the time taken for one cycle and is shown in Fig. 2.5.

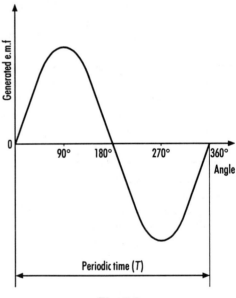

Fig. 2.5

The frequency of an a.c. waveform is defined as the number of cycles it completes per second, and is measured in hertz (Hz). This means that one cycle per second equals 1 Hz. Therefore if the coil rotates twice per second, the frequency is

2 Hz. In the UK the frequency of our a.c. supply is 50 Hz. The name hertz may sound strange but it is the name of a German scientist.

The relationship between periodic time and frequency (f) is $T = 1/f$ so that $f = 1/T$.

2.5 ANGULAR VELOCITY

Look again at Fig. 2.3(a). The coil is moving in an anticlockwise direction and in a circular movement. This movement is given the name angular velocity (ω) and is measured in radians per second. You may remember that the circumference of a circle is given by $2\pi r$ where r is the radius of the circle. Now from this we can state that there are 2π radians in one complete cycle. Finally angular velocity $= 2\pi\times$ frequency or

$$\omega = 2\pi f$$

Figure 2.6 shows one cycle of an a.c. sine wave plotted to a base of time and angular measure.

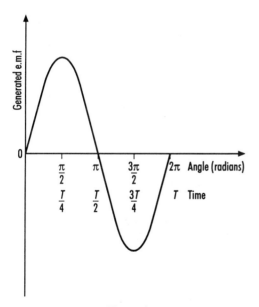

Fig. 2.6

2.6 INSTANTANEOUS VOLTAGE AND CURRENT

Look again at Fig. 2.3(a). This time look at the position of the angle θ. At any instant in time the angle θ turned through in time (t) seconds equals the angular velocity times the time taken, i.e.

$$\theta = \omega t$$

From Section 2.3 we know that $v = V \sin \theta$. So

$$v = V \sin \omega t$$

Likewise, the instantaneous current, i, is given by

$$i = I \sin \omega t$$

From Section 2.5 we know that $\omega = 2\pi f$ so finally we can state that

$$v = V \sin 2\pi f t$$

and

$$i = I \sin 2\pi f t$$

Examples

1. Determine the frequency for a period of 0.4 seconds (s).

$$T = 0.4\,s$$

Using $f = 1/T$

$$f = \frac{1}{0.4} = 2.5\,Hz$$

2. What is the periodic time for a frequency of 50 Hz?

$$f = 50\,Hz$$

Using $T = 1/f$

$$T = \frac{1}{50} = 0.2\,s$$

3. A sinusoidal alternating voltage has a maximum value of 100 V and a frequency of 500 Hz. Write down an expression for the instantaneous voltage and calculate this value at an instant 0.6 milliseconds (ms) after the start of a cycle.
   Sketch a sine wave showing the position of the calculated instantaneous voltage.

$$V = 100\,V \quad f = 500\,Hz$$
$$t = 0.6\,ms = 0.0006\,s$$

(Always try and remember to convert the milliseconds to seconds.)

Using $v = V \sin 2\pi f t$ at $0.0006\,\text{s}$ after the start of the cycle

$$v = 100 \sin (2\pi \times 500 \times 0.0006)$$

$$= 100 \sin (1.885)$$

At this point it is essential to remember that the 1.885 is in radian measure. So to calculate $\sin 1.885$ you must set your calculator to radian measure. Do this now and check my value for $\sin 1.885$.

$$v = 100 \times 0.951 = 95.1\,\text{V}$$

Figure 2.7 shows the position of the instantaneous voltage after a time of $0.6\,\text{ms}$.

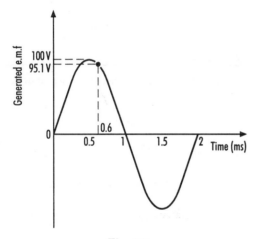

Fig. 2.7

4. The instantaneous value of an a.c. is given by $i = 5 \sin 314.2t$. Determine the current after 12 ms and sketch a sine wave showing the position of the current.

$$1000 \text{ ms} = 1\,\text{s}$$

so

$$t = 12\,\text{ms} = 0.012\,\text{s}$$

Using $i = 5 \sin 314.2t$

$$i = 5 \sin (314.2 \times 0.012)$$

$$= 5 \sin (3.7704 \text{ radians})$$

$$= 5 \times -0.5882 = -2.941\,\text{A}$$

Figure 2.8 shows the position of the current after a time of 12 ms.

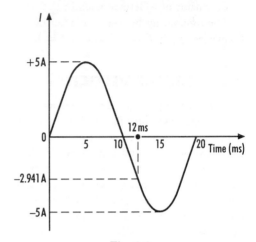

Fig. 2.8

This example is to remind you that an alternating waveform changes from a positive half cycle to a negative half cycle and in this case the time of 12 ms places the current in the negative half with a value of 2.941 A.

5. A sinusoidal a.c. is given by $i = 50 \sin 628.4t$. Determine: (a) the peak value; (b) the frequency; (c) the periodic time.

    (a) Look carefully at the general equation $i = I \sin 2\pi f t$. Now write down the given equation

    $$i = 50 \sin 628.4t$$

    You should notice that I and the 50 are linked. This means that the peak value $I = 50\,\text{A}$.

    (b) Also you should notice that $2\pi f t$ are linked to $628.4t$. Therefore

    $$2\pi f t = 628.4t$$

    so

    $$f = \frac{628.4t}{2\pi t}$$

    The t on the top of the calculation and the t at the bottom cancel out, leaving

    $$f = \frac{628.4}{2\pi} = 100\,\text{Hz}$$

(c) Periodic time $T = 1/f$. So

$$T = \frac{1}{50} = 0.02\,s = 20\,ms$$

Exercise 2.1

1. Determine the frequency for a period of 0.2 s.

2. What is the periodic time for a quantity which has a frequency of 100 Hz?

3. Define the term 'frequency'.

4. An a.c. is represented by the equation $i = 10\sin 314.2t$. Determine the frequency and the instantaneous current after 3 ms and 14 ms. Sketch a waveform showing the position of each instantaneous current.

5. A sinusoidal voltage waveform has a frequency of 50 Hz and a peak value of 20 V. Plot a graph showing the variation of voltage with time for one cycle of the waveform.

6. Calculate the periodic time corresponding to the frequencies of: (a) 40 Hz; (b) 2 kHz; (c) 5 MHz.

7. Calculate the frequencies corresponding to periods of: (a) 0.02 s; (b) 4×10^{-4} s; (c) 5 ms; (d) 8 μs.

8. A sinusoidal voltage has a peak value of 35 V and an angular velocity of 1000 rad/s. Express this voltage in the form $v = V\sin 2\pi ft$.

9. Determine the frequency and periodic time for an alternating e.m.f. having an instantaneous value of 5 V, 0.02 s after passing through zero if the peak value is 12 V.

10. The current produced by an alternating e.m.f. has a peak value of 30 A. Determine the instantaneous value of the current when the loop is 40° anticlockwise from the horizontal.

11. An alternating e.m.f. is represented by $v = 25\sin\theta$. Determine the value of v when θ equals: (a) 30°; (b) 60°; (c) 90°; (d) 180°; (e) 210°; (f) 270°.

12. Draw to scale one complete cycle of a sine wave of current with a maximum value of 5 mA and a frequency of 2000 Hz. From the graph, determine the instantaneous current at: (a) 0.1; (b) 0.3; (c) 0.4; (d) 0.6 ms from the time when the current passed through zero and was increasing in value.

13. A 50 Hz sinusoidal current has a maximum value of 10 A. Plot this current wave over one complete cycle to a base of time.
    From the graph, find the times when the instantaneous current value is 5 A.

14. Draw a sine wave of current having a maximum value of 10 A to the following scale: 10 mm $\equiv$ 1 A, 30 mm $\equiv$ 30°. Find from the graph the root mean square (r.m.s.) value of the current.

15. Explain the meaning of the following terms as applied to an alternating quantity: (a) instantaneous value; (b) peak value; (c) frequency.

2.7 PEAK-TO-PEAK VALUES OF VOLTAGE

In the case of a pure alternating waveform we can see from Fig. 2.9 that the peak voltage equals the maximum voltage and that the peak-to-peak voltage is twice the maximum voltage.

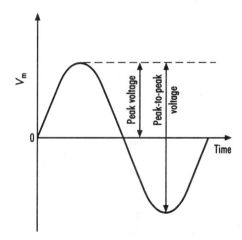

Fig. 2.9

If the fluctuating voltage never falls below zero, the peak voltage will be equal to or greater than the peak-to-peak voltage. This situation is illustrated in Fig. 2.10.

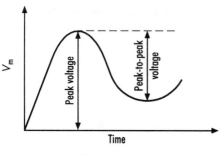

Fig. 2.10

2.8 AVERAGE OR MEAN VALUE OF AN ALTERNATING WAVEFORM

General rule – to find the average of a set of quantities, add the quantities together and divide by the number of quantities in the set.

The average value of the sinusoidal wave illustrated in Fig. 2.11 is zero because half of the area is positive, above the time axis, and half is negative, below the axis.

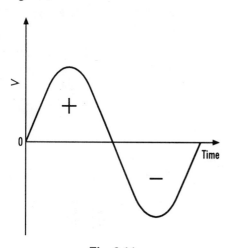

Fig. 2.11

In electrical engineering the average value is only found for one half cycle of a wave.

Examples

1. Find the average value of the current (I_{av}) for the waveform illustrated in the figure.

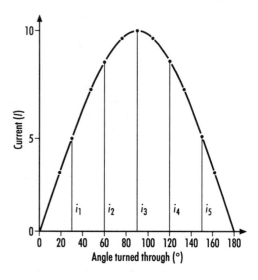

$$I_{av} = \frac{i_1 + i_2 + i_3 + i_4 + i_5}{5}$$

$$= \frac{5 + 8.66 + 10 + 8.66 + 5}{5}$$

$$= \frac{37.32}{5} = 7.464 \, A$$

The above is an approximate method. The average value can be found accurately by using the mid-ordinate rule

$$\text{average value} = \frac{\text{sum of mid-ordinates}}{\text{number of mid-ordinates}}$$

2. Determine the average value of current (I_{av}) over one half cycle using the mid-ordinate rule for the waveform illustrated below.

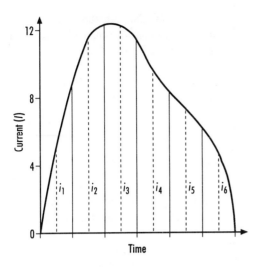

$$I_{av} = \frac{i_1 + i_2 + i_3 + i_4 + i_5 + i_6}{6}$$

$$= \frac{5 + 11.6 + 12.4 + 9.2 + 7.2 + 4.8}{6}$$

$$= \frac{50.2}{6} = 8.37\,\text{A}$$

The general rule

$$I_{av} = \frac{i_1 + i_2 + \cdots + i_n}{n}$$

applies to all symmetrical waveforms taken over one half cycle.

By experiment it can be shown that for a sine wave $I_{av} = 0.637 \times I_m$ (the maximum current), likewise $V_{av} = 0.637 \times V_m$.

2.9 ROOT MEAN SQUARE VALUE OF AN ALTERNATING WAVEFORM

The r.m.s. values are the effective values of an alternating quantity that are registered on ammeters or voltmeters when connected in an electrical circuit.

The general rule is

$$I = \sqrt{\left(\frac{i_1^2 + i_2^2 + \cdots + i_n^2}{n}\right)}$$

By experiment it can be shown that for a sine wave

$$I = 0.707 I_m$$

and

$$V = 0.707 V_m$$

In electrical engineering unless otherwise stated the quoted values are r.m.s. values.

Examples

1. Determine the r.m.s. value for the waveform in the figure.

$$I = \sqrt{\left(\frac{5^2 + 11.6^2 + 12.4^2 + 9.2^2 + 7.2^2 + 4.8^2}{6}\right)}$$

$$= \sqrt{\left(\frac{25 + 134.56 + 153.76 + 84.64 + 51.84 + 23.04}{6}\right)}$$

$$= \sqrt{\left(\frac{472.84}{6}\right)} = \sqrt{78.81} = 8.878\,\text{A}$$

2. A sinusoidal a.c. has a maximum value of 10 A. What is: (a) the r.m.s. value; (b) the average value?

$$I_m = 10\,\text{A}$$

(a) For a sine wave

$$I = 0.707 I_m = 0.707 \times 10 = 7.07\,\text{A}$$

(b) For a sine wave

$$I_{av} = 0.637 I_m = 0.637 \times 10 = 6.37\,\text{A}$$

3. The half-wave of an alternating e.m.f. has the following values at the times shown.

E.m.f. (V)	0	10	40	60	55	45	12	0
Time (s)	0	0.1	0.3	0.5	0.6	0.7	0.9	1

Plot a graph and if the points are joined by straight lines determine: (a) the average value; (b) the r.m.s. value.

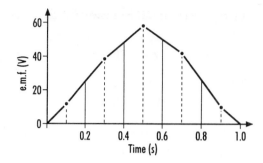

(a)

(b)

$$V_{av} = \frac{10 + 40 + 60 + 45 + 12}{5}$$

$$= \frac{167}{5} = 33.4\,V$$

(b)

$$V_{r.m.s.} = \sqrt{\left(\frac{10^2 + 40^2 + 60^2 + 45^2 + 12^2}{5}\right)}$$

$$= \sqrt{\left(\frac{100 + 1600 + 3600 + 2025 + 144}{5}\right)}$$

$$= \sqrt{\left(\frac{7469}{5}\right)} = \sqrt{1493.8} = 38.64\,V$$

Exercise 2.2

1. Determine the peak value, average value, r.m.s. value, frequency, and periodic time of a current $i = 4.3 \sin 200\pi t$.

2. The voltage across a resistor is given by $v = 10 \sin 628.4t$. Calculate: (a) the frequency of supply; (b) the periodic time; (c) the r.m.s. value.

3. Draw to scale the sine wave representing a current having an amplitude of 25 A. By means of the mid-ordinate rule find: (a) the average value; (b) the r.m.s. value.

4. Draw to scale a sinusoidal waveform having a peak value of 100 V. By means of the mid-ordinate rule, determine: (a) average value; (b) r.m.s. value.

5. Plot graphically two sinusoidal alternating voltages of peak values 30 V and 40 V respectively, the former leading by 90°.

Determine the peak and r.m.s. values of their resultant and the angle by which it lags behind the 30 V waveform.

6. A stepped a.c. wave has the following values over equal intervals of time:

Value	4	6	6	4	2	0	0	−2	−4
Time intervals	0–1	1–2	2–3	3–4	4–5	5–6	6–7	7–8	8–9

What value of d.c. would give the same heating effect as this waveform?

7. What is meant by the r.m.s. value of an alternating voltage, and why is this value used?

An alternating voltage varies over one half-cycle as follows:

Time (s)	0	0.001	0.002	0.003	0.004	0.005	0.006
Voltage	0	8.5	17	25	17	8.5	0

Plot the curve and deduce the r.m.s. value of the voltage.

8. What is meant by the r.m.s. value of an a.c. and why is this value used?

A sinusoidal a.c. has a maximum value of 2 A. Draw to scale the curve of current over half a cycle and obtain graphically the r.m.s. value of the current.

9. The time between the positive peak of an alternating voltage wave and the first succeeding negative peak is 0.02 s. What is the frequency? Plot this voltage wave and obtain the r.m.s. value from your graph, assuming the wave to be sinusoidal and 200 V peak wave. If this voltage is applied to a 57.6 Ω resistor, calculate the power in watts.

10. What is understood by the r.m.s. value of an a.c.? A sinusoidal current has a maximum value of 5 A. Plot the current to scale and determine graphically its r.m.s. value. If the current flows for 15 minutes through a 20 Ω resistor, what energy, in joules, is dissipated in the resistor?

11. What is meant by the r.m.s. value of an a.c.? How is its numerical value determined graphically?

A sinusoidal current has a maximum value of 56.56 A. How many commercial units of electrical energy (kW h) would be dissipated if it flowed through a 5 Ω resistor for 2 hours?

12. An a.c. had the following values for one half-cycle:

Angle (rad)	0	$\pi/9$	$\pi/6$	$\pi/3$	$\pi/2$	$2\pi/3$	$5\pi/6$	$8\pi/9$	π
Current (A)	0	5	20	35	40	35	20	5	0

These current values are joined by straight lines. Obtain the r.m.s. value of this current from your graph. If the time of one complete cycle is 0.015 s, what is the frequency?

13. Explain the meaning of the following terms as applied to an alternating quantity: (a) instantaneous value; (b) frequency.

An a.c. waveform is given by $i = 4 \sin 314t$. Plot this waveform to a base of time (in seconds) and thence show, by graphical means, that the r.m.s. value is 2.83 A.

14. An alternating voltage has a stepped waveform which varies as follows for equal intervals of time: 2, 3, 4, 2, 1, −2, −3, −4, −2, −1 V. Plot this and thence determine for the complete wave the r.m.s. value.

15. What is meant by the r.m.s. value of an a.c.? Explain why this value is used. An alternating current has the following values over one-half of a cycle:

Time (ms)	0	1	2	3	4	5	6	7	8	9	10
Current (A)	0	2.5	3.4	3.1	3.3	3.6	3.3	3.1	3.4	2.5	0

Determine the r.m.s. value and the frequency of the current wave.

2.10 PHASE DIFFERENCE

Alternating voltages and currents vary in value from instant to instant. Most circuits have this voltage and current, so it is important to consider their relationships.

2.10.1 In-phase

When two alternating quantities have the same frequency and have positive peaks which occur at the same time, as shown in Fig. 2.12, they are said to be in-phase.

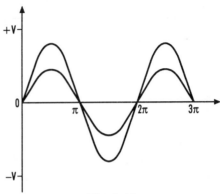

Fig. 2.12

2.10.2 Leading

In Fig. 2.13 v_2 is leading v_1. The clues to look for when making a decision are: (a) check the position of each peak value – if they are not occurring at the same time then there is a phase difference; (b) check that one of the quantities is starting from zero time – the other quantity will be above or below the axes. If it is above, the quantity will be leading.

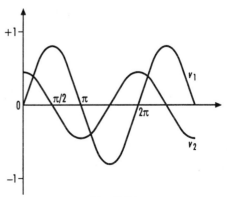

Fig. 2.13

2.10.3 Lagging

In Fig. 2.14 v_2 is lagging v_1. In this case the peaks are not occurring at the same time so there is a phase difference. One quantity is starting from zero. The other quantity is below the axes which means that v_2 is lagging v_1.

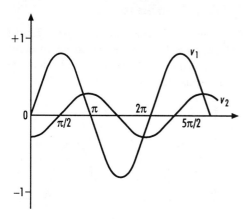

Fig. 2.14

We have considered only relationships between v_1 and v_2 but it is important to remember that we can consider other relationships. For example, in a particular circuit we might consider two currents or in another circuit currents and voltages.

2.11 PHASE ANGLE

The current shown in Fig. 2.15 leads the voltage by $\pi/2$ radians (90°). This angle is called the phase angle ($\phi°$). The general formula introduced in Section 2.4 can now be modified to take into account the phase angle. In this case the formula becomes $i = I \sin (\omega t + \phi)$ and $v = V \sin \omega t$.

Examples

1. State the phase relationship between $v = 10 \sin \omega t$ and $i = 4 \sin [\omega t + (\pi/4)]$. By inspection:

    $$i \text{ leads } v \text{ by } \frac{\pi}{4} \text{ radians } (45°)$$

2. State the relationship between $v = 24 \sin \omega t$ and $i = \sin[\omega t - (\pi/3)]$. By inspection:

    $$i \text{ lags } v \text{ by } \frac{\pi}{3} \text{ radians } (60°)$$

3. State the relationship between the voltage v_1 and v_2 and determine the instantaneous voltages of v_1 and v_2 at a time of 3 ms.

    $$v_1 = 10 \sin 314.2t$$

    and

    $$v_2 = 8 \sin \left(314.2t + \frac{\pi}{3} \right)$$

    $t = 3 \text{ ms} = 3 \times 10^{-3}$ s. Therefore

    $$v_1 = 10 \sin 314.2t$$
    $$= 10 \sin (314.2 \times 3 \times 10^{-3})$$
    $$= 10 \sin 0.9426$$
    $$= 10 \times 0.8091 = 8.091 \text{ V}$$

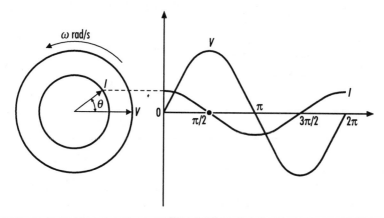

Fig. 2.15

and

$$v_2 = 8 \sin\left(314.2t + \frac{\pi}{3}\right)$$

$$= 8 \sin(0.9426 + 1.0472)$$

$$= 8 \sin 1.9898$$

$$= 8 \times 0.9135 = 7.31\,\text{V}$$

v_2 leads v_1 by $\pi/3$ radians, i.e. 60°.

Exercise 2.3

1. A waveform is given by $v = V \sin(2\pi ft + \phi)$. State the quantity that each symbol represents.

2. Draw one cycle of the waveform $v = 10 \sin 314.2t$.

3. Draw waveforms showing the relationship between $v = 10 \sin \omega t$ and $i = 15 \sin \omega t$.

4. State the relationship between i and v and draw the two waveforms when $v = 100 \sin 2\pi ft$ and $i = 20 \sin[(2\pi ft + (\pi/3)]$.

5. Two a.c.s are represented by $i_1 = 20 \sin \omega t$ and $i_2 = 30 \sin[\omega t + (\pi/5)]$. Draw one cycle of each waveform.

6. Write down the phase relationship between the two quantities in each of the following pairs:

   (a) $v_1 = V_1 \sin \omega t$;
   $v_2 = V_2 \sin(\omega t + \phi)$
   (b) $v_1 = V_1 \sin \omega t$;
   $v_2 = V_2 \sin(\omega t - \phi)$
   (c) $v_1 = V_1 \sin \omega t$;
   $v_2 = V_2 \sin(\omega t + \pi)$
   (d) $v_1 = V_1 \sin \omega t$;
   $v_2 = V_2 \sin[\omega t - (\pi/2)]$
   (e) $v_1 = 10 \sin \theta$;
   $v_2 = 5 \sin[\theta + (\pi/3)]$
   (f) $v_1 = 5 \sin \theta$;
   $v_2 = 12 \sin[\theta - (\pi/2)]$
   (g) $v_1 = 25 \sin \omega t$;
   $v_2 = 100 \sin[\omega t - (\pi/4)]$
   (h) $v_1 = 12 \sin \omega t$;
   $v_2 = 20 \sin[\omega t + (\pi/3)]$

   (i) $i_1 = 8 \sin \omega t$;
   $i_2 = 10 \sin[\omega t + (\pi/8)]$
   (j) $i_1 = 20 \sin \omega t$;
   $i_2 = 10 \sin[\omega t - (\pi/4)]$
   (k) $v = 100 \sin \omega t$;
   $i = 25 \sin[\omega t + (\pi/5)]$
   (l) $v = 25 \sin \theta$;
   $i = 4 \sin[\theta - (\pi/6)]$

7. Calculate the instantaneous voltages from $v_1 = 15 \sin 314.2t$ and $v_2 = 25 \sin[314.2t + (\pi/3)]$ where $t = 4$ ms. State the phase relationship between these voltages.

8. Determine the maximum value, the r.m.s. value, and the instantaneous value for a time of 2 ms of the following voltages: $v_1 = 20 \sin 314.2t$ and $v_2 = 10 \sin[314.2t + (\pi/2)]$. State the phase relationship between the voltages.

9. A voltage given by $v = 50 \sin 314.2t$ is applied to a circuit which produces a current $i = 25 \sin[314.2t - (\pi/4)]$. Determine: (a) the instantaneous values of v and i for a time of 3 ms; (b) the r.m.s. value of the current; (c) the frequency of the supply; (d) the phase angle of the current relative to the voltage.

10. (a) Explain the term r.m.s. value as applied to an a.c. (b) An a.c. flowing through a circuit has a maximum value of 70 A and lags the applied voltage by 60°. The maximum value of the voltage is 100 V and both current and voltage waveforms are sinusoidal.

    Plot the current and voltage waveforms in their correct relationships for the positive half of the voltage. What is the value of the current when the voltage is at a positive peak?

11. Two sinusoidal currents separated by a phase angle of 60° and having maximum values of 6 A and 8 A (the smaller leading in phase) are fed into one conductor. Plot these waveforms and hence determine the maximum value of the resultant current and give its phase relationship with respect to the larger current.

12. Define the terms r.m.s. value and average value as applied to an a.c. and state the relationship of each to the peak value for a current which varies sinusoidally.

Draw a complete cycle of a sinusoidal voltage of peak value 6 V, using a voltage scale of 10 mm ≡ 1 V. On the same diagram, show, to the same scales, a second sinusoidal voltage of peak value 3 V and leading the first voltage by a phase angle of 45°.

Answers————————————

Exercise 2.1

1. 5 Hz
2. 0.01 s
4. 50 Hz, 8.09 A, −9.5 A
6. (a) 0.025 s; (b) 0.5 ms; (c) 0.2 μs
7. (a) 50 Hz; (b) 2500 Hz; (c) 200 Hz; (d) 125 000 Hz
8. $V = 35 \sin 1000t$
9. 3.419 Hz, 0.2925 s
10. 19.28 A
11. (a) 12.5 V; (b) 21.65 V; (c) 25 V; (d) 0; (e) −12.5 V; (f) −25 V
12. (a) 4.756 mA; (b) −2.939 mA; (c) −4.756 mA; (d) +4.756 mA
13. 0.001 67 s, 0.008 33 s
14. 7.07 A

Exercise 2.2

1. 4.3 A, 2.739 A, 3.04 A, 100 Hz, 10 ms
2. (a) 100 Hz; (b) 0.01 s; (c) 7.07 V
3. (a) 15.93 A; (b) 17.68 A
4. 63.7 V; 70.7 V
5. 50 V, 35.4 V, 53°08′
6. 3.74 A
7. 14.5 V
8. 1.414 A
9. 25 Hz, 141.4 V, 347 W
10. 3.535 A, 225 kJ
11. 16 kW h
12. 27.3 A, 66.7 Hz
14. 2.6 V
15. 2.93 A, 50 Hz

Exercise 2.3

6. (a) v_2 leads v_1 by ϕ; (b) v_2 lags v_1 by ϕ; (c) v_2 leads v_1 by 180°; (d) v_2 lags v_1 by 90°; (e) v_2 leads v_1 by 60°; (f) v_2 lags v_1 by 90°; (g) v_2 lags v_1 by 45°; (h) v_2 leads v_1 by 60°; (i) i_2 leads i_1 by 22.5°; (j) i_2 lags i_1 by 45°; (k) i leads v by 36°; (l) i lags v by 30°
7. 18.55 V, v_2 leads v_1 by 60°
8. $v_{1m} = 20$, $v_{1r.m.s.} = 14.14$, $v_1 = 11.76$, $v_{2m} = 10$ V, $v_{2r.m.s.} = 7.07$ V, $v_2 = 8.09$ V
9. (a) 40.96 V; (b) 3.93 A; (c) 50 Hz; (d) 45° lagging
10. 35 A
11. 12.18 A, 25°16′

3

WAVEFORMS AND SIGNALS

3.1 DIRECT AND ALTERNATING CURRENT

3.1.1 Direct current

Figure 3.1 shows a waveform for a continuous or steady d.c. and Fig. 3.2 shows a waveform with a varying d.c. Both of these waveforms are unidirectional because they flow in one direction only. The waveforms do not cross the time axis.

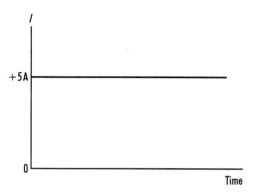

Fig. 3.1

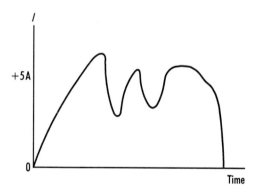

Fig. 3.2

A new battery or a smoothed power pack would produce the output shown in Fig. 3.1 and a power pack with an unsmoothed output would produce the waveform shape in Fig. 3.2.

3.1.2 Alternating current

A generator or alternator produces an a.c. In the a.c. waveform of Fig. 3.3, which is a sine wave, the current rises from zero to a maximum value in one positive direction, and then falls to zero, and then becomes maximum again in a negative direction before returning to zero.

This sine wave shape is the waveform produced by a power station to provide household supplies at 230 $V_{\text{r.m.s.}}$ 50 Hz. The peak value will be

$$\frac{V_{\text{r.m.s.}}}{0.707} = \frac{230}{0.707} = 325\,\text{V}$$

Other a.c. waveforms are shown in Fig. 3.4.

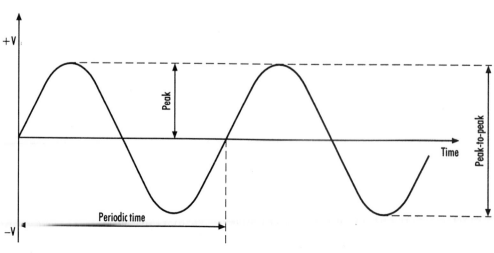

Fig. 3.3

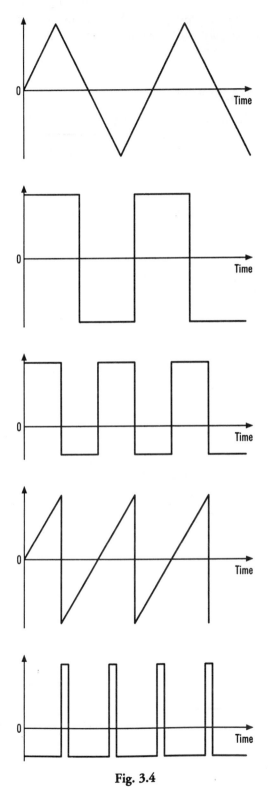

Fig. 3.4

3.2 SIGNALS

In electronics an electrical quantity that varies with time so as to convey information is said to be a signal. When using equipment, the signals are often in the form of varying voltages or currents.

3.2.1 Signal processing

This is the operation which must be carried out on a signal to achieve the desired end result. Typical operations are waveform shaping, amplification, frequency changing etc.

3.2.2 Signal generator

A signal generator can provide various types of waveform used to test circuits and systems. There are many makes available, so ask your teacher to show you one available in your workroom. The signal generator consists of an alternating voltage generator called an oscillator followed by a potential divider so that the amplitude of a signal can be varied. There are two basic types of signal generator: (a) audio frequency generator; and (b) radio frequency generator.

An audio frequency generator covers the range from about 0.1 Hz to about 100 kHz whereas the radio frequency generator covers the range from 100 kHz to about 450 MHz.

In general terms, signal generators produce a sine wave output and a square wave output. If other types of waveform are needed then we can use a function generator which will provide such waveforms as triangular etc.

3.2.3 Analogue signal

The main property of an analogue signal can have any value and may be the amplitude, phase or frequency of an electronic signal, the angular position of a shaft or the pressure of a gas or fluid. An audio, i.e. sound, signal is an example of an analogue signal. Signals are often called analogue when it is necessary to compare them to digital signals.

3.2.4 Digital signal

The main property of a digital signal can have only a limited number of discrete values. Discrete values means distinct values as compared to several values. The term is widely used in binary transmissions where there are only two discrete values.

A useful example of this is when you walk into a dark room at home and operate the light switch. You are using a two-state device. The switch is either ON or OFF. When it is OFF you are using 0 V and when it is ON you are using 230 V to supply a lamp. To simplify matters we can say that we are in a '0' or '1' situation.

3.3 WAVEFORM TERMS

Whenever waveforms are being dealt with you will come across, time and time again, names used to describe them. The following names deal with these terms.

3.3.1 Leading and trailing edge

Figure 3.5 shows a waveform – the arrows show which edge is which.

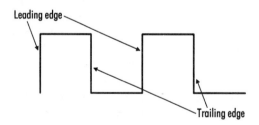

Fig. 3.5

3.3.2 Pulse

The terms pulse or pulsed can be applied to any waveform which is not a true sine wave as shown in Fig. 3.6.

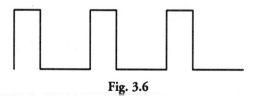

Fig. 3.6

However, they are usually used to describe rectangular waveforms which can also include square pulses.

3.3.3 Period

This term has been dealt with before, but it fits nicely into this section so it is repeated. Figure 3.7 shows the periodic time which is the time taken to complete one cycle of the waveform.

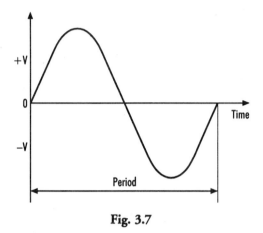

Fig. 3.7

3.3.4 Pulse width or pulse duration

This is the length of time (t) in each period during which the waveform departs from its steady state. Look carefully at the different positions of the steady state in Figs 3.8(a) and (b).

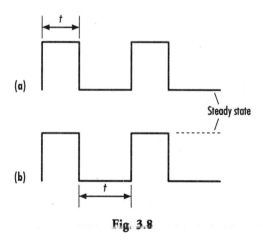

Fig. 3.8

3.3.5 Mark-to-space ratio

Figure 3.9 shows a rectangular waveform. The time the pulse operates at one level is called the 'mark' and the time the pulse operates at the other level is called the 'space'. The mark-to-space ratio shows how much faster the switching process of one of the oscillator switches operates compared to the other. In the case of a square wave the mark-to-space ratio is unity because the mark is equal to the space.

$$\text{mark-to-space ratio} = \frac{t_1}{t_2}$$

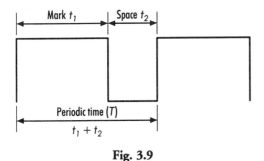

Fig. 3.9

3.3.6 Pulse duty factor or duty ratio

The duty factor is defined as the ratio of the pulse duration to the pulse period. From Fig. 3.9

$$\text{pulse duty factor} = \frac{t_1}{t_1 + t_2}$$

3.3.7 Pulse repetition frequency

The pulse repetition frequency is the number of pulses occurring in one second. From Fig. 3.9:

$$\text{pulse repetition frequency} = \frac{1}{t_1 + t_2}$$

Example

1. A rectangular waveform has a mark time (t_1) of 0.75 s and a space time (t_2) of 0.5 s with an amplitude of 5 V. Calculate: (a) mark-to-space ratio; (b) pulse duty factor; (c) pulse repetition frequency.

$$t_1 = 0.75 \text{ s} \quad t_2 = 0.5 \text{ s}$$

(a) Mark-to-space ratio is

$$\frac{t_1}{t_2} = \frac{0.75}{0.5} = 1.5$$

(b) Pulse duty factor is

$$\frac{t_1}{t_1 + t_2} = \frac{0.75}{0.75 + 0.5} = \frac{0.75}{1.25} = 0.6$$

(c) Pulse repetition frequency is

$$\frac{1}{t_1 + t_2} = \frac{1}{0.75 + 0.5}$$

$$= \frac{1}{1.25} = 0.8 \text{ Hz}$$

Exercise 3.1

1. Explain with the aid of sketches the following terms: (a) steady d.c.; (b) varying d.c.

2. Draw an a.c. waveform for one complete cycle. Label the voltage and time axes. Show the peak value and the periodic time.

3. Explain what is meant by the following terms: (a) signal; (b) signal processing; (c) analogue signal; (d) digital signal.

4. Explain, with the aid of sketches, what is meant by the following terms: (a) leading and trailing edge; (b) pulse; (c) period; (d) pulse duration.

5. Draw a labelled diagram of a rectangular waveform that has a mark of 2 s and a space of 0.5 s with an amplitude of 5 V. Calculate: (a) mark-to-space ratio; (b) pulse duty factor; (c) pulse repetition frequency.

6. From the trace in the figure if the amplitude control is set to 0.4 V per division and the base control is set to 50 μs per division determine: (a) amplitude; (b) periodic time; (c) frequency; (d) mark-to-space ratio.

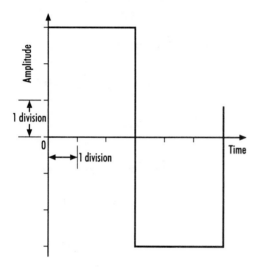

7. Match the numbers on the waveform in the figure with the following terms: (a) space; (b) mark; (c) amplitude; (d) periodic time.

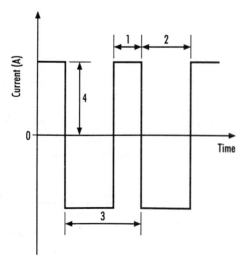

8. A rectangular waveform varying between 0 V and +12 V has the following characteristics: duration of positive pulse 3 μs; pulse repetition frequency 100 kHz. Determine: (a) mark-to-space ratio; (b) pulse duty factor.

9. Draw a square wave having a pulse repetition frequency of 2 MHz. What is the mark-to-space ratio and the pulse duty factor of this wave?

10. Draw to scale the following waveforms: (a) sine wave with a peak of 3 V and a periodic time of 12 ms; (b) square wave of amplitude 5 V and a periodic time of 10 ms; (c) rectangular wave of maximum value 4 V, mark time of 1 s and a space time of 0.2 s.

3.4 INTEGRATING CIRCUIT

The RC circuit shown in Fig. 3.10 is known as an integrating circuit. You need not remember this name but you should study carefully the output waveforms produced under various conditions for values of R and C shown in Figs 3.10 and 3.11.

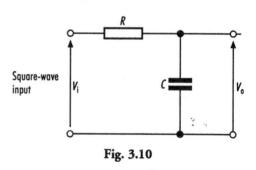

Fig. 3.10

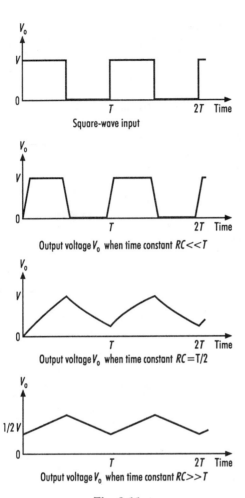

Square-wave input

Output voltage V_o when time constant $RC \ll T$

Output voltage V_o when time constant $RC = T/2$

Output voltage V_o when time constant $RC \gg T$

Fig. 3.11

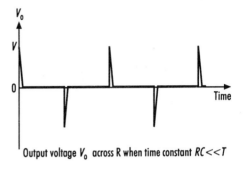

Square-wave input

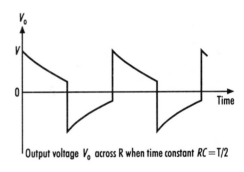

Output voltage V_o across R when time constant $RC \ll T$

Output voltage V_o across R when time constant $RC = T/2$

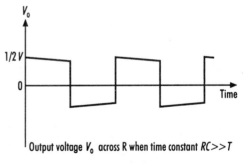

Output voltage V_o across R when time constant $RC \gg T$

Fig. 3.13

3.5 DIFFERENTIATING CIRCUIT

The resistor and capacitor of the previous circuit when interchanged produce a differentiating circuit as shown in Fig. 3.12. Again, the name is not important but you need to study the effect of different values of RC on the output waveforms as shown in Fig. 3.13.

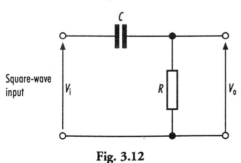

Fig. 3.12

3.6 RECTIFIER CIRCUITS

Electrical energy is distributed in an a.c. form but many industrial applications need d.c. The basic input and output waveforms provided by using a smoothed rectifier circuit are shown in Fig. 3.14.

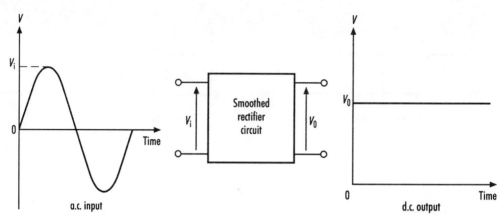

Fig. 3.14

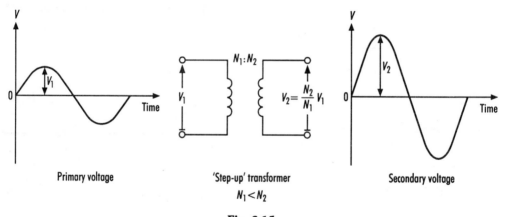

Fig. 3.15

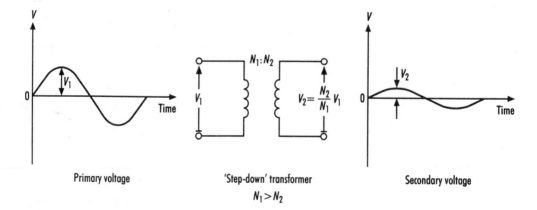

Fig. 3.16

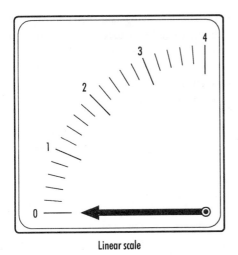 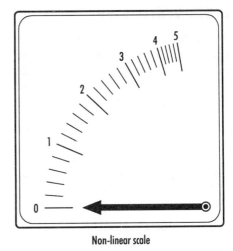

Linear scale   Non-linear scale

Fig. 3.17

3.7 TRANSFORMER

In rectifier circuits it is often the case that the input voltage is applied via a transformer. Figure 3.15 shows the case for a step-up transformer. The secondary output voltage is greater than the input primary voltage.

Figure 3.16 shows the alternative application, that of a step-down transformer. In this case the output voltage is smaller than the input voltage.

3.8 MEASURING INSTRUMENTS

In electronics, it is necessary to measure aspects of the signals that have been mentioned in this chapter. We also need to exhibit on screen the shapes that can be obtained from various circuits. To take these measurements we need instruments.

- *Analogue-type instrument*   this is the type of instrument that uses a pointer and scale. The pointer moves over a graduated scale. The scale can either be linear or non-linear as shown in Fig. 3.17.
- *Digital-type instrument*   this type of instrument is widely in use because it gives a direct read-out in decimal numbers. We can obtain instruments using digital displays to measure current (a.c. or d.c.), voltage (a.c. or d.c.), frequency, temperature etc.

The following is an overview of what is available. In later chapters you will be given the opportu-

nity to carry out investigations using instruments.

3.8.1 Ammeter

An instrument for the measurement of current; it is available in analogue or digital form. The ammeter is always connected in series with the circuit component under test. Care should be taken to connect the ammeter the correct way round otherwise in an analogue instrument the pointer will be deflected in the wrong direction.

3.8.2 Voltmeter

An instrument for measuring the potential difference between two points. The voltmeter is always connected across i.e. in parallel with the circuit component under test. Voltmeters are available in analogue or digital form. A circuit diagram using an ammeter and a voltmeter is shown in Fig. 3.18.

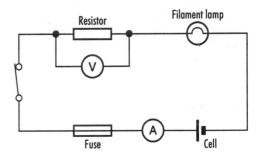

Fig. 3.18

3.8.3 Wattmeter

An instrument for measuring power on both a.c. and d.c. circuits. Figure 3.19 shows the connections for measuring power supplied to a load. The basic wattmeter is called an electrodynamic instrument.

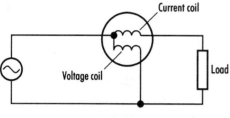

Fig. 3.19

The current coil is fixed and the voltage coil moves as shown in Fig. 3.20. The current in the moving coil is proportional to the voltage while the load current is passed through the fixed coils. The turning movement is proportional to the mean product of voltage and current.

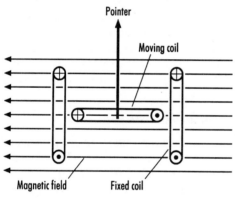

Fig. 3.20

The reason a wattmeter can be used on a.c. without a rectifier is that when the direction of the current in the moving coil is reversed the current in the fixed coil is also reversed and the turning moment remains in the same direction.

3.8.4 Ohmmeter

An instrument that is used to determine the resistance of a circuit or a circuit component. The instrument contains a built-in source of

e.m.f. and a moving coil meter movement in addition to resistors used for calibration. The principle of operation is:

1. short the test leads together;
2. adjust the built-in variable resistance until the meter reads full scale;
3. separate the test leads and connect them to the unknown resistance;
4. read off the resistance value.

In an analogue instrument the ohmmeter scale reads in the opposite direction to other instruments, i.e. the zero reading is full-scale deflection. A general layout is shown in Fig. 3.21.

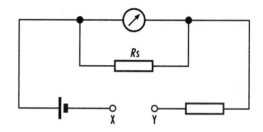

Fig. 3.21

3.8.5 Multimeter

A multimeter is an instrument that incorporates a voltmeter, ammeter, and ohmmeter and can therefore be set to measure either voltage, current or resistance. Both a.c. and d.c. ranges are available.

3.8.6 Cathode-ray oscilloscope

The cathode-ray oscilloscope (CRO) is an instrument that enables a variety of electrical signals to be observed and examined visually. Applied voltages are indicated by the deflection of a spot of light on the face of the tube. The spot records instantaneous values and can therefore be used to display a.c. waveforms when moved from left to right in a series of scans by a time base of adjustable frequency. A typical simple CRO is illustrated in Fig. 3.22. A set of squares is marked on the CRO screen to help with measurements.

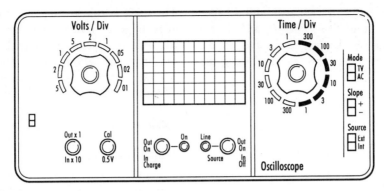

Fig. 3.22

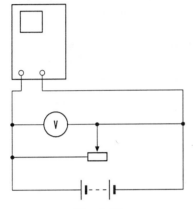

Fig. 3.23

Measurement of d.c. voltage

The circuit is set up with the components as illustrated in Fig. 3.23.

The CRO is switched on an allowed to warm up. Once the display has been established, the voltmeter reading and the deflection on the graticule can be noted. By varying the resistance, several sets of readings can be taken and a calibration graph plotted as illustrated in Fig. 3.24.

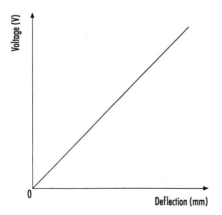

Fig. 3.24

The battery is now replaced with the source of unknown voltage. The CRO is switched on, the deflection noted, and the CRO switched off. From the calibration graph the value of the unknown cell can be determined.

Measurement of d.c.

The circuit is set up as illustrated in Fig. 3.25 with the unknown current I passing through the known resistor R.

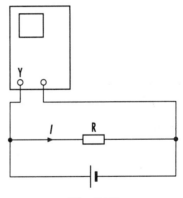

Fig. 3.25

The potential difference across R is determined by using the d.c. voltage calibration graph plotted in the last experiment. The value of the current I is calculated using

$$I = \frac{V}{R}$$

Measurement of resistance

With the Y input switched off, the CRO controls are adjusted until the spot is central. The circuit is set up as illustrated in Fig. 3.26.

Resistors R_1 and R_2 are adjusted until the spot returns to centre. The values R_1 and R_2 are noted. With R_3 known, the value of the unknown resistor can be calculated using the relationship

$$R = \frac{R_1 R_2}{R_3}$$

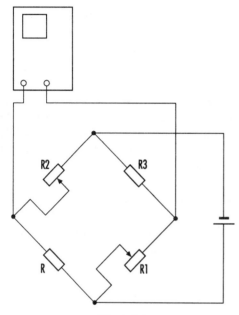

Fig. 3.26

Measurement of a.c. voltage
The circuit is connected up as illustrated in Fig. 3.27.

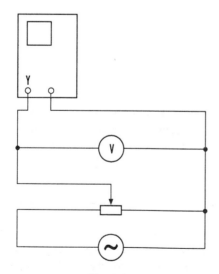

Fig. 3.27

The CRO is switched on and at least six sets of readings are taken in the same way as the d.c. voltage experiment. A calibration graph similar to Fig. 3.24 is plotted. The a.c. supply is now replaced by the unknown e.m.f. The CRO is switched on, the deflection noted, and from the calibration graph the value of the unknown e.m.f. is determined.

Measurement of a.c.
The method is similar to that of the previous experiment and that of the experiment used to determine the value of a d.c.

Measurement of frequency
The circuit is set up as illustrated in Fig. 3.28 with a calibrated time base.

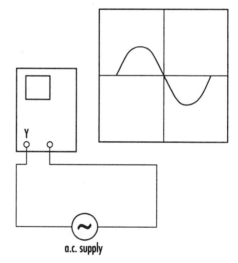

a.c. supply

Fig. 3.28

The CRO and signal generator are switched on and allowed to warm up. The frequency of the time-base is adjusted until one complete cycle appears on the screen. The frequency can then be read from the dial of the time-base generator.

Measurement of frequency using Lissajous figures
The principle of frequency measurement using Lissajous figures is by means of comparing the waveform of unknown frequency with a waveform of known frequency.

The circuit is set up as illustrated in Fig. 3.29 with the internal time-base generator of the CRO switched off.

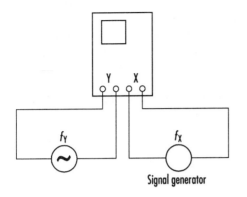

Fig. 3.29

With the known frequency f_Y applied to the Y plates the unknown frequency f_X is applied to the X plates. Adjustments are made until one of the Lissajous figures is illustrated as in Fig. 3.30.

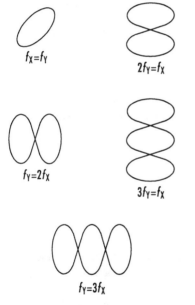

Fig. 3.30

The frequency to be measured is calculated from

$$\frac{f_X}{f_Y} = \frac{\text{number of horizontal loops}}{\text{number of vertical loops}}$$

When using this method of frequency measurement it is advisable to start with $f_X = f_Y$

because this gives a single loop and subsequent counting errors are likely to be reduced. The accuracy of this method depends upon the accuracy of the known frequency f_Y.

Measurement of phase difference

The phase difference for the circuit illustrated in Fig. 3.31 can be displayed on a double-beam CRO as shown by the waveform. This method is suitable only when a rough value is required. A more accurate method is by using the circuit of Fig. 3.29. The phase difference is determined by reference to Fig. 3.32.

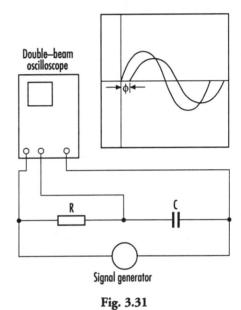

Fig. 3.31

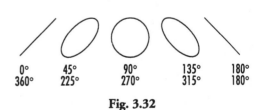

Fig. 3.32

Examples

1. Sketch the display that would appear on the screen of a CRO if an alternating voltage is applied to the Y plates with the time-base (a) off, and (b) on.

(a) With the time-base off this means that the spot will only move up and down as shown in the figure.

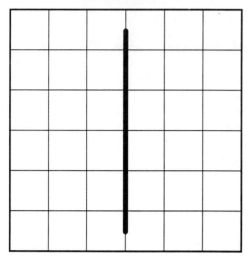

(b) With the time-base on this means that the Y plates allow the spot to move up and down; and at the same time with the X plates on, the spot will sweep across the screen. The result of both of these movements is shown below.

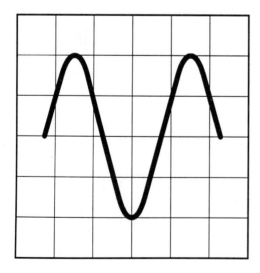

2. Figure (a), at the top of the facing page, shows an a.c. supply connected to a resistor. The trace obtained on the CRO connected to the resistor is shown in (b). The vertical sensitivity is set to 2 V per division and the time-base is set to 0.5 ms per division. For the signal shown calculate: (a) time for the signal to cross the screen once; (b) periodic time; (c) frequency; (d) peak value; (e) average value; (f) r.m.s. value.

(a) Time (t) for the signal to cross the screen once = 9 divisions $\times$ 0.5 ms per division. Therefore

$$t = 9 \times 0.5 = 4.5 \, \text{ms}$$

(b) Period time (T) = time for one cycle which is 6 divisions $\times$ 0.5 ms per division. So

$$T = 6 \times 0.5 = 3 \, \text{ms} = 0.003 \, \text{s}$$

(c) Frequency $(f) = 1/T$. So

$$f = \frac{1}{0.003} = 333.3 \, \text{Hz}$$

(d) Peak voltage (V) = 3 divisions $\times$ 2 V per division. Therefore

$$V = 3 \times 2 = 6 \, \text{V}$$

(e) Average value $(V_{av}) = 0.637 \times$ peak value and so

$$V_{av} = 0.637 \times 6 = 3.822 \, \text{V}$$

(f) Root mean square value $(V_{r.m.s.}) = 0.707 \times$ peak value. Therefore

$$V_{r.m.s.} = 0.707 \times 6 = 4.242 \, \text{V}$$

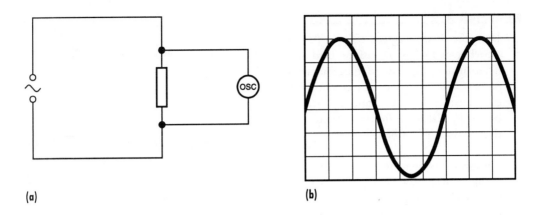

(a)

(b)

Example 2

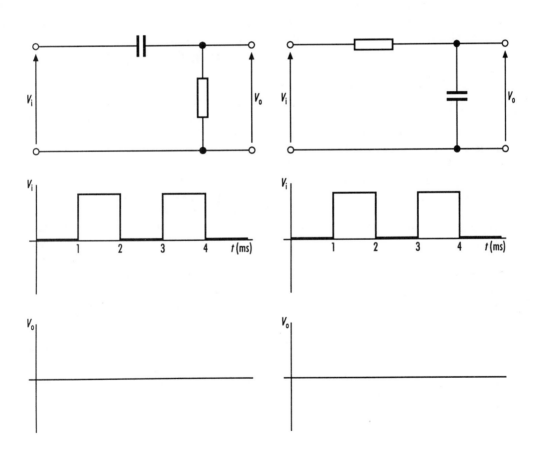

Exercise 3.2(2)

Exercise 3.2

1. Draw the circuit diagram of: (a) a 330 pF capacitor and a 56 kΩ resistor used as a differentiator; (b) a 0.22 μF capacitor and a 360 kΩ resistor used as an integrator. For each circuit identify the input and output terminals. Calculate for each circuit the time constant (T) where $T = RC$.

2. The figure (bottom of p. 39) shows two circuits each with an input waveform. Copy out the circuit diagrams and input waveforms and underneath the waveforms draw the output waveforms.

3. A 230 V 50 Hz supply is the input voltage to the following systems: (a) step-down transformer; (b) step-up transformer. For each case, draw the input waveform and the shape you would expect for the output wave-form.

4. State the main difference between an analogue and a digital type measuring instrument.

5. Draw a circuit diagram showing a cell, a single pole switch, a fuse, a resistor and an ammeter all connected in series. Include in your diagram a voltmeter to measure the cell e.m.f.

6. Part of a circuit diagram is shown below. Name the type of instrument that is being used in each position.

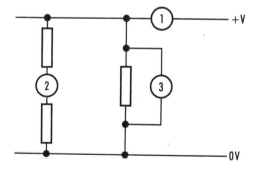

7. The figure shows the scale of a multimeter set to 500 mA full-scale deflection. State the value of current that is being indicated by the pointer.

8. The figure shows the scale of a multimeter set to 400 V full-scale deflection. State the value of voltage that is being indicated by the pointer.

9. Draw a waveform for each of the following cases: (a) speech from a microphone; (b) digital signal in a computer.

10. Draw a circuit diagram showing a watt-meter being used to measure the power taken by a load resistor.

11. The figure shows a trace on a CRO. The vertical sensitivity of the oscilloscope is set to 2 V per division and the time-base is set to 0.4 ms per division. Calculate the following quantities: (a) peak value of the signal; (b) r.m.s. value of the signal; (c) average value of the signal; (d) time taken for the signal to cross the screen once; (e) periodic time; (f) frequency. State one practical use of a CRO.

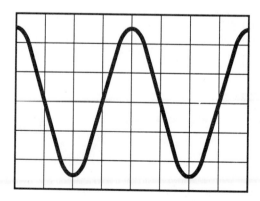

12. Draw a grid of the same dimensions as that in question 11. The vertical sensitivity is set to 10 V per division and the time-base is set to 1 ms per division. Plot on it a sine wave peak value 20 V and periodic time 6 ms. Calculate: (a) time taken to cross the screen once; (b) frequency; (c) average value of the signal; (d) r.m.s. value of the signal.

13. Draw a grid of the same dimensions as that in question 11. Draw on it a rectangular wave of amplitude 6 V with a mark time of 40 ms and a space time of 20 ms. The vertical sensitivity is set to 2 V per division and the time-base is set to 10 ms per division. Determine: (a) periodic time; (b) mark-to-space ratio; (c) pulse duty factor; (d) pulse repetition frequency.

14. Describe, with the aid of a circuit diagram, how an oscilloscope would be used to take the following measurements: (a) measurement of d.c. voltage; (b) measurement of d.c.; (c) measurement of resistance; (d) measurement of frequency using Lissajous figures; (e) measurement of phase difference.

15. Explain with the aid of a sketch how an ohmmeter can be used to measure the resistance of a component.

Answers

Exercise 3.1
1. See 3.1
2. See Fig. 3.3
3. See 3.2, 3.2.1, 3.2.3, 3.2.4
4. See 3.3
5. See Fig. 3.9 (a) 4; (b) 0.8; (c) 0.4
6. (a) 1.2 V; (b) 0.0003 s; (c) 3333 Hz; (d) 1:1
7. (a) 2; (b) 1; (c) 4; (d) 3
8. (a) 3:7; (b) 0.3
9. (a) 1:1; (b) 0.5
10. See Figs 3.7, 3.5, 3.9

Exercise 3.2
1. (a) See Fig. 3.12, 18.48 μs; (b) see Fig. 3.10, 79.2 ms
2. (a) Differentiator circuit, see Fig. 3.12; (b) Integrator circuit, see Fig. 3.10
3. (a) See Fig. 3.16; (b) see Fig. 3.15
4. Analogue instrument uses a scale and pointer. Digital instrument uses a decimal readout.
5. See Fig. 3.18
6. 1 – ammeter; 2 – ammeter; 3 – voltmeter.
7. 300 mA
8. 240 V
9. (a) Similar to Fig. 3.2; (b) see Fig. 3.6
10. See Fig. 3.19
11. (a) 5 V; (b) 3.535 V; (c) 3.185 V; (d) 3.2 ms; (e) 1.6 ms; (f) 625 Hz
12. See figure in Q.11. (a) 8 ms; (b) 166.7 Hz; (c) 12.74 V; (d) 14.14 V
13. (a) 60 ms; (b) 2:1; (c) 0.67; (d) 16.67 Hz
14. See 3.6
15. See 3.4

4

COMPONENTS 1

4.1 RESISTORS

.A resistor is a component used in a circuit because it has the property of electrical resistance. Resistance is defined as the opposition to a flow of electrons in an electrical circuit. It is very important because it is the property which determines the amount of current that will flow in an electrical circuit. Current is equal to the potential difference at the terminals of a resistor divided by the resistance, i.e.

$$\text{current} = \frac{\text{potential difference}}{\text{resistance}}$$

or

$$I = \frac{V}{R}$$

Calculations using this formula have been dealt with in Chapter 1.

Resistors are manufactured in many ways and provide a wide range of characteristics. A few are considered.

4.1.1 Carbon-composition resistor

In general terms this is a low cost and common form of resistor. This resistor is made from carbon and a mixture of other materials moulded into shape under pressure. These other materials are included so that different resistance values can be obtained by varying the amount of carbon present.

The rating for this type of resistor is generally given as its total resistance and power rating. Physically these resistors are very small and it is difficult to print on them the actual resistance

value. Instead a colour code is used to show the value of the resistance and its tolerance.

Figure 4.1 shows the general shape and typical dimensions of a carbon composition resistor.

Tolerance means that, for example, a $100\,\Omega$ resistor can range in value from $90\,\Omega$ to $110\,\Omega$ if the tolerance was $\pm10\%$. The closer you get to the resistor nominal value the more you pay for the resistor.

Manufacturers can make resistors to any nominal value as long as the customer is prepared to pay the asking price. However, many manufacturers provide resistors in a range of what is called preferred values. This means that they only make certain values. Two series of resistors are used, E12 for 10% tolerance and E24 for 5% tolerance. More about this in Section 4.3.

One problem with this type of resistor is that it has poor long-term stability. Stability means that regardless of working conditions the resistance value should not change significantly. Under some conditions, such as temperature variation, hot or cold conditions, high operating voltages, dampness etc., the resistance value does change.

Example

1. A $5\,\Omega$ resistor rated at $0.25\,\text{W}$ is being considered for use in a circuit with a supply of $2\,\text{V}$. Determine whether the resistor has a suitable power rating.

$$R = 5\,\Omega \quad V = 2\,\text{V} \quad P = 0.25\,\text{W}$$

Using $I = V/R$

$$I = \tfrac{2}{5} = 0.4\,\text{A}$$

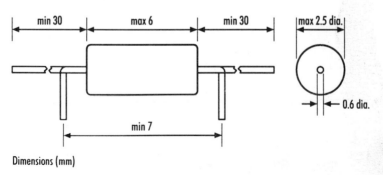

min 30   max 6   min 30 |max 2.5 dia.|

min 7

0.6 dia.

Dimensions (mm)

Fig. 4.1

Using $P = VI$

$$P = 2 \times 0.4 = 0.8\,\text{W}$$

Because the resistor is rated at 0.25 W and the actual power consumption will be 0.8 W, the resistor is **unsuitable**.

4.1.2 Wire-wound resistors

This type of resistor is manufactured by winding a length of wire on a former. Several different types of wire are used, the more common being nichrome, constantan and manganin. The final product tends to be strong, an all-welded unit with ceramic formers, vitreous enamelled or silicone coated. They are generally very reliable and can be used for high power applications. Power ratings range from less than 1 W to several kW. Physically, as you would probably expect, they are larger than carbon-composition resistors. By nature of their design they are strong so that they can withstand the frequent start–stop cycle characteristics of motor starting and other similar applications. Some wire-wound resistors have a ribbed construction; this helps when they get hot and rapid cooling is needed.

Wire-wound resistors are generally not used above a frequency of 50 kHz because they have undesirable inductive and capacitive effects on the circuit in use. Applications for wire-wound resistors are as load resistors, shunt resistors, voltage divider networks, filament dropping resistors, voltage dropping resistors, grid resistors, hi-volt bleeder resistors in power supplies, bias supply resistors, etc.

Wire-wound resistors are often protected by a vitreous enamel coating which is flame resistant. The markings on the resistor can therefore resist high temperatures, solvents and abrasions.

Most wire-wound resistors have solder-coated radial lug terminals which are suitable for soldered or bolt connections into a circuit.

4.1.3 Carbon-film resistor

In this type of resistor a thin film of hard, crystalline carbon is deposited on a ceramic or glass former. As compared to a carbon-

composition-type resistor they have a higher degree of stability.

Physically they look very much like the carbon-composition resistor with a typical body diameter of 2.5 mm and length 6 mm. A typical technical specification would be:

Power rating	0.25 W at 70°C
Resistance range	1 Ω to 10 MΩ
Voltage rating	250 V
Maximum overload voltage	600 V
Temperature range	−25 to +70°C

Most manufacturers provide other information but it is important that when choosing a resistor you make your selection very carefully considering all of the points mentioned to satisfy your circuit requirements.

4.1.4 Metal-film resistors

Thick film surface-mounted resistors

As the name suggests they are mounted on the surface and not by wire terminals as with a carbon-composition resistor. This type of resistor offers higher performance than carbon film with very low noise levels and high reliability.

Figure 4.2 shows in cross-section a typical thick film resistor. The multiple-layer construction with nickel barrier termination offers a very good surface for any soldering that needs to take place.

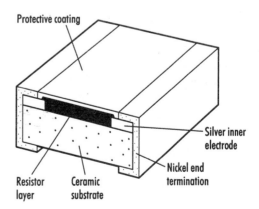

Fig. 4.2

Precision metal-film resistors

Physically these resistors look like carbon composition resistors. Figure 4.3 shows typical

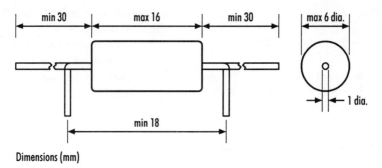

Dimensions (mm)

Fig. 4.3

dimensions. The resistance range available is from $1\,\Omega$ to $10\,M\Omega$ with a maximum power rating of about 1.5 W. The temperature range is usually from -55 to $+155°C$ with a maximum working voltage of about 500 V. Some manufacturers provide a low-inductance resistor of this type for use with higher frequency applications. Because these resistors are precision types they are usually given a special exterior coating for extra environmental protection.

4.2 THE BRITISH STANDARD 1852 CODE

When reading a circuit diagram it is important to be able to read the values of the resistors in use. This is a code that can be used to do this. The capital letter R is placed in the decimal point position. Capital K denotes kilo and capital M denotes mega. The following examples show how it is used.

R22	$0.22\,\Omega$
1R0	$1\,\Omega$
2R2	$2.2\,\Omega$
22R	$22\,\Omega$
220R	$220\,\Omega$
2K0	$2\,k\Omega$
2K2	$2.2\,k\Omega$
22K0	$22\,k\Omega$
2M2	$2.2\,M\Omega$
22M0	$22\,M\Omega$

In addition to this code there is a code for tolerance. The letters used are $F = \pm1\%$, $G = \pm2\%$, $J = \pm5\%$, $K = \pm10\%$, $M = +20\%$.

Care needs to be taken not to confuse the tolerance letters K and M with the resistor value letters.

Example

1. Write down the code for a $470\,\Omega$ resistor with a tolerance of 5%.

$$470\Omega \pm 5\% \quad \text{is } 470RJ$$

4.3 PREFERRED VALUES OF RESISTORS

To save on manufacturing costs and to prevent having too many resistor values available only certain values are made. These resistor values are called preferred values. There are two series in general use, the E12 for 10% tolerance and the E24 for 5% tolerance.

4.3.1 E12 series

As you might expect there are 12 values available. They are:

10 12 15 18 22 27 33 39 47 56 68 82

4.3.2 E24 series

The values shown are in addition to those in the E12 series. In total this will make available 24 values:

11 13 16 20 24 30 36 43 51 62 75 91

Examples

1. From the E12 series write down a number of resistors that are available using the base number 47.

   The resistors available are R47, 4R7, 47R, 470R, 4K7, 47K, 470K, 4M7, 47M, 470M.

2. From the E24 series write down the value of $3.6\,\Omega \pm 10\%$ using the 1852 code.

   A resistor of value $3.6\,\Omega \pm 10\%$ is written down as 3R6K.

3. From the E12 series write down the value of $560\,\Omega \pm 5\%$ using the 1852 code and calculate the minimum and maximum value that the resistor can have.

   A resistor of value $560\,\Omega \pm 5\%$ is written down as 560RJ. The minimum value is

   $$560 - 5\% \text{ of } 560 = 560 - (0.05 \times 560)$$
   $$= 560 - 28 = 532\,\Omega$$

   The maximum value is

   $$560 + 5\% \text{ of } 560 = 560 + (0.05 \times 560)$$
   $$= 560 + 28 = 588\,\Omega$$

4.4 RESISTOR COLOUR CODE

It is easier if resistors are marked with coloured bands rather than numbers because of their physical size. A colour coded resistor looks like the sketch shown in Fig. 4.4. A, B, C and D are the coloured bands. A and B represent the first two numbers of the resistance, C is the number of zeros and the fourth band is the tolerance of the resistor. The colours used to represent the different numbers are shown in Table 4.1.

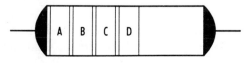

Fig. 4.4

Table 4.1

Black	0	Violet	7
Brown	1	Grey	8
Red	2	White	9
Orange	3	Gold	$\pm 5\%$
Yellow	4	Silver	$\pm 10\%$
Green	5	No colour	$\pm 20\%$
Blue	6		

Examples

1. Write down the value of a resistor colour coded Brown, Black, Black, Gold.

   Brown indicates 1
   Black indicates 0
   Black indicates that there are no zeros
   Gold indicates $\pm 5\%$

   The resistor value is 10 ohms $\pm 5\%$.

2. Write down the colour code for a resistor of value $120\,\Omega \pm 20\%$.

   First number indicates Brown
   Second number indicates Red
   Third number indicates number of zeros, i.e. one, which is coded Brown

   The colour code for $120\,\Omega$ is Brown, Red, Brown. There is no fourth band.

3. Give the values of these resistors in Ω and $k\Omega$: (a) Brown, Grey, Red; (b) Red, Red, Red; (c) Red, Violet, Orange, Gold; (d) Orange, White, Yellow, Silver.

   (a) Brown is 1, Grey is 8, Red is 2.
       Resistor value is $1800\,\Omega \pm 20\%$ or $1.8\,k\Omega \pm 20\%$.
   (b) Red is 2, Red is 2, Red is 2.
       Resistor value is $2200\,\Omega \pm 20\%$ or $2.2\,k\Omega \pm 20\%$.
   (c) Red is 2, Violet is 7, Orange is 3, Gold is $\pm 5\%$.
       Resistor value is $27\,000\,\Omega \pm 5\%$ or $27\,k\Omega \pm 5\%$.
   (d) Orange is 3, White is 9, Yellow is 4, Silver is $\pm 10\%$.
       Resistor value is $390\,000\,\Omega \pm 10\%$ or $390\,k\Omega \pm 10\%$.

4. Write down the colour code you would expect to see on the following resistors: (a) 16 Ω; (b) 240 Ω; (c) 3.6 kΩ ± 5%; (d) 620 kΩ ± 10%; (e) 43 MΩ.

(a) Brown is 1, Blue is 6, Black for no zeros, No colour is ±20%.
Colour code is Brown, Blue, Black.

(b) Red is 2, Yellow is 4, one zero is Brown.
Colour code is Red, Yellow, Brown.

(c) 3.6 kΩ = 3600 Ω
Orange is 3, Blue is 6, Two zeros is Red, ±5% is Gold.
Colour code is Orange, Blue, Red, Gold.

(d) 620 kΩ = 620 000 Ω
Blue is 6, Red is 2, four zeros is Yellow, ±10% is silver
Colour code is Blue, Red, Yellow, Silver.

(e) 43 MΩ = 43 000 000 Ω
Yellow is 4, Orange is 3, six zeros is Blue, No colour is 20% tolerance.
Colour code is Yellow, Orange, Blue.

Exercise 4.1 _____

1. Explain what is meant by the terms resistance and resistor.

2. Describe, with the aid of sketches, the construction and action of the following: (a) carbon-composition resistor; (b) wire-wound resistor; (c) carbon-film resistor; (d) metal-film resistor.

3. A 100 Ω resistor rated at 0.5 W is being considered for use in a circuit with a supply of 6 V. Determine whether the resistor has a suitable power rating.

4. A 12 Ω resistor is rated at 1 W. Determine whether the resistor can be safely used in a circuit with a supply of 6 V.

5. Copy out the following table and determine in each case whether the resistor can be used safely under the given conditions.

Suggested resistor	Potential difference	Current	Power
100 Ω 1 W	2 V		
36 Ω 2 W	6 V		
250 Ω 4 W	10 V		
500 Ω 10 W	230 V		
15 Ω 0.5 W	6 V		
180 Ω 0.25 W	4 V		

6. State the value of a resistor that is coded 470RJ.

7. A resistor has a value 510 Ω ± 10%. Write this out in terms of the British Standard 1852 code.

8. State the value of the following resistors that have been coded according to the 1852 code: (a) R82; (b) 2R0; (c) 37R; (d) 160R; (e) 1K8; (f) 5K6; (g) 2M2; (h) 56RJ; (i) 820RK; (j) 3K9J; (k) 8K2M; (l) 2M4K.

9. Write out the following resistor values using the British Standard 1852 code: (a) 0.18 Ω; (b) 1 Ω; (c) 5.6 Ω; (d) 20 Ω; (e) 220 Ω; (f) 1.5 kΩ; (g) 33 kΩ; (h) 5.6 MΩ; (i) 51 Ω ± 5%; (j) 620 Ω ± 10%; (k) 750 Ω ± 20%.

10. What is the colour code for the following resistor values? (a) 15 Ω ± 5%; (b) 100 Ω ± 10%; (c) 3.9 kΩ ± 20%; (d) 12 kΩ ± 5%; (e) 220 kΩ ± 10%; (f) 6.8 MΩ ± 20%.

11. Write down the resistance of each of the following resistors: (a) Brown, Black, Black; (b) Brown, Black, Brown; (c) Yellow, Violet, Brown; (d) Red, Black, Orange; (e) Red, Red, Red; (f) Orange, Orange, Red; (g) Brown, Black, Orange; (h) Yellow, Violet, Yellow; (i) Violet, Green, Black, Silver; (j) Blue, Grey, Brown, Gold; (k) White, Brown, Red, Gold; (l) Orange, White, Black, Silver.

12. What E12 preferred values would you use if you calculated that a circuit needed resistors having values of: (a) $11\,\Omega$; (b) $130\,\Omega$; (c) $170\,\Omega$; (d) $1900\,\Omega$; (e) $21\,\text{k}\Omega$; (f) $290\,\text{k}\Omega$; (g) $4.6\,\text{M}\Omega$; (h) $75\,\text{M}\Omega$.

13. Calculate the maximum and minimum value a resistor might have if it is nominally rated at $200\,\Omega \pm 10\%$.

14. For the following resistors calculate the highest and lowest values possible of resistance: (a) $100\,\Omega \pm 5\%$; (b) $120\,\Omega \pm 10\%$; (c) $150\,\Omega \pm 20\%$; (d) $1.6\,\text{k}\Omega \pm 5\%$; (e) $1.8\,\text{k}\Omega \pm 10\%$; (f) $20\,\text{k}\Omega \pm 20\%$; (g) $240\,\text{k}\Omega \pm 5\%$; (h) $300\,\text{k}\Omega \pm 10\%$; (i) $360\,\text{k}\Omega \pm 20\%$; (j) $4.3\,\text{M}\Omega \pm 5\%$; (k) $5.6\,\text{M}\Omega \pm 10\%$; (l) $6.2\,\text{M}\Omega \pm 20\%$; (m) $68\,\text{M}\Omega \pm 5\%$; (n) $75\,\text{M}\Omega \pm 10\%$; (o) $82\,\text{M}\Omega \pm 20\%$.

15. Write down the 1852 code for a resistor of value $160\,\Omega \pm 10\%$ and calculate the minimum and maximum value that the resistor can have.

16. Complete the following table for maximum and minimum value of resistors.

Resistor value (Ω)	5% tolerance		10% tolerance		20% tolerance	
	max.	min.	max.	min.	max.	min.
10						
120						
1500						
18						
220						
2700						
33						
390						

4.5 POTENTIAL OR VOLTAGE DIVIDER

4.5.1 Basic circuit

Figure 4.5 shows two resistors connected in series. The input voltage is shown connected across both resistors with the output voltage coming from one resistor. The input voltage is therefore divided.

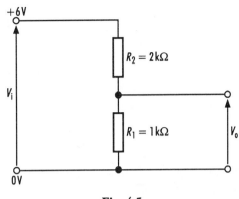

Fig. 4.5

For a potential divider the output voltage (V_o) is given by:

$$V_o = \frac{\text{signal resistance } (R_1) \times \text{input voltage } (V_i)}{\text{total resistance } (R_T)}$$

where $R_T = R_1 + R_2$.

Example

1. For the circuit shown in Fig. 4.5 calculate the output voltage.

$$R_1 = 1000\,\Omega$$

$$R_T = 1000 + 2000 = 3000\,\Omega$$

Using

$$V_o = (R_1 \times V_i)/R_T$$

$$V_o = \frac{1000 \times 6}{3000} = 2\,\text{V}$$

Potential dividers can have more than two resistors connected in series to give you a wide range of output voltages.

4.5.2 Commercial voltage divider

A typical metal-film precision voltage divider network is shown in Fig. 4.6. The resistance range available is very wide, from $50\,\Omega$ to

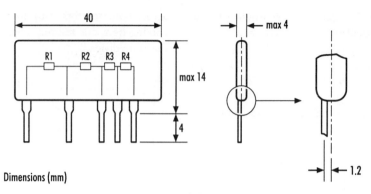

Dimensions (mm)

Fig. 4.6

10 MΩ, with a low tolerance figure of less than 1%. Maximum working voltages are often up to 1200 V a.c. or d.c. This type of voltage divider has many practical applications, in particular for use in summing amplifiers, input switching in digital voltmeters etc.

4.6 VARIABLE RESISTOR (VR)

A VR, as its name suggests, is a type of resistor where the resistance can be varied. The British Standard graphical symbol is shown in Fig. 4.7.

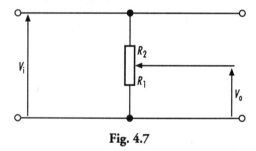

Fig. 4.7

When all three connections are used it will act as a potential divider. The output voltage can be varied by changing the ratio of resistance R_1 to R_2. A number of different types are available.

4.6.1 Slider-type VR

A length of wire is wound on to a former and each end of the wire has a terminal. A slider is mounted on a bar above the resistance wire, making contact with the wire. A third terminal is

mounted at one end of the sliding bar. To change the resistance ratio the sliding bar is moved along the track until the desired position is reached.

4.6.2 Lug-type wire-wound adjustable resistor

Figure 4.8 shows a typical wire-wound adjustable resistor. Resistance wire is wound on to a former and coated with vitreous enamel but a narrow strip is left uncoated to expose a portion of each turn of the resistance wire. Contact to any wire on this exposed portion is made by an adjustable metal lug. The lug is adjusted by using a screwdriver. The lug is slackened off, moved to its desired position, tightened up, and then connected in the circuit. This type of adjustable resistor can have quite high power ratings, for example up to 25 W.

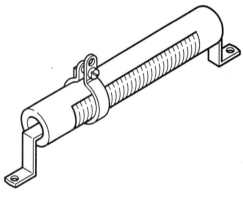

Fig. 4.8

4.6.3 Wire-wound power rheostats

Another name for a VR is a rheostat. Figure 4.9 shows a typical rheostat. The resistance wire is wound on a solid ceramic former with a terminal at each end. The third terminal is a contact brush made from metal graphite. When used in free air, the power rating can go up to 300 W.

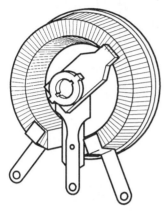

Fig. 4.9

4.6.4 Enclosed-type VR

The VR is totally enclosed and sealed against moisture and other undesirable environmental elements. Typically they are about 25 mm in diameter and 15 mm deep. As you would now expect in a VR there are three terminals. The centre terminal is connected to the rotating contact to give a variable resistance depending upon its position on the wire resistor circular track. A typical enclosed VR is shown in Fig. 4.10.

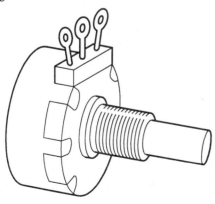

Fig. 4.10

This type of VR has a wide range of resistance values up to $100\,k\Omega$ with a power rating of about 2 W. The operating temperature range is from $-55°C$ to $+120°C$ with a working voltage of up to 500 V. The rotating central contact is designed for a rotational life of at least 25 000 cycles.

Sometimes this type of VR is called a 'pot' because of the term '**pot**ential divider'. This is quite in order but it is unfortunate that they are also called potentiometers. This term is technically incorrect. However, it is used by manufacturers and is here to stay. So be careful with your choice of technical terms when dealing with such matters.

A typical application of this type of VR is in volume control of virtually any type of sound system on the market.

4.6.5 Skeleton-type pre-set trimmer

A typical pre-set trimmer is shown in Fig. 4.11. The circular track material is either carbon or cermet. The word cermet comes from the beginning of **cer**amic and the beginning of **met**al oxide. They are designed for easy setting, long life and stability. Physically they are smaller than the enclosed-type VR and have a lower power rating such as about 0.5 W.

4.7 CONTROL OF CURRENT AND VOLTAGE

There are a number of ways in which a VR can be used.

4.7.1 Control of current

Figure 4.12 shows a circuit diagram of a VR being used to control current in a lighting circuit. Practical applications of this can be found in the home when a dimmer switch is used to lower the intensity of the light in a room and when the lighting in a cinema or theatre is controlled. When the sliding contact of the VR is moved towards the right the resistance is increased and the current is reduced. In a

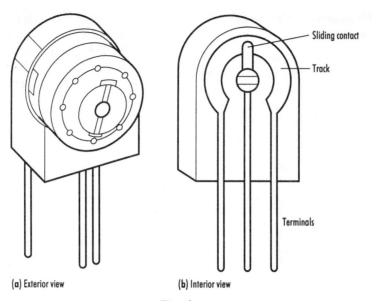

(a) Exterior view   (b) Interior view

Fig. 4.11

similar manner, if the sliding contact is moved towards the left, the resistance is reduced and the current increased.

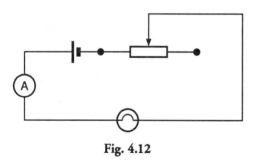

Fig. 4.12

An ammeter is connected in series with the lamp to measure current flow. For accuracy it is essential to use a good quality ammeter. Such an ammeter will have negligible resistance and therefore will not significantly add to the total circuit resistance.

4.7.2 Control of voltage

Figure 4.13 shows a circuit diagram of a VR being used to control voltage. This circuit layout should remind you of a potential divider. With a potential divider the values of resistance are fixed. With a VR they can be changed so as to give a range of outputs. When a voltmeter is connected across the output it is useful to have

one with a very high resistance. In this case only a small amount of current will go through it when it is connected in a circuit.

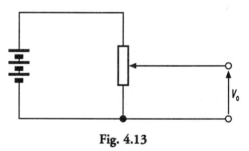

Fig. 4.13

When the sliding contact of the VR is moved down the output voltage will be relatively low and when the slider is moved upwards the output voltage is relatively high.

The two resistors in a potential divider and the top and bottom resistors in a VR are given names. The top resistor is called the pull-up resistor and the bottom one is called the pull-down resistor. So, if you pull the slider down, down comes the output voltage. In a like manner, if you pull the slider up, up comes the output voltage.

Examples_____

1. A VR of total resistance (15) kΩ is connected in series with a resistor of value

100 Ω. The input voltage is 6 V. If the VR is set in the ratio top:bottom as 2:1 draw a circuit diagram and determine the output voltage. Explain the need for the 100 Ω resistor.

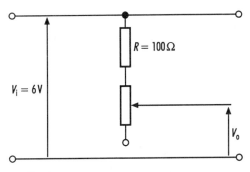

Resistors,

$$\text{Top:Bottom} = 15 \text{ k}\Omega$$
$$2:1 = 15 \text{ k}\Omega$$

logically $10:5 = 15 \text{ k}\Omega$

therefore $R_1 = 5 \text{ k}\Omega$

and

$$V_o = \frac{R_1 \times V_i}{R_T}$$

$$= \frac{5 \times 6}{15} = 2 \text{ V}$$

The 100 Ω resistor is in the circuit so that if by any chance the VR is set to its theoretical minimum value of zero there will always be some current present so as not to cause any problems. In the calculation we have ignored the 100 Ω resistor.

2. Consider the circuit in the figure above. Calculate the current (I) flowing through the 100 Ω resistor if the VR is set to: (a) its minimum value; (b) its maximum value.

(a) VR minimum value $= 0 \Omega$ and $R = 0 + 100 = 100 \Omega$. Using $I = V_i/R$

$$I = \frac{6}{100} = 0.06 \text{ A}$$

(b) VR   maximum   value $= 15 \text{ k}\Omega = 15\,000 \Omega$ and $R = 15\,000 + 100 = 15\,100 \Omega$. Using $I = V_i/R$

$$I = \frac{6}{15\,100} = 0.0\,003\,974 \text{ A}$$

4.8 CURRENT CONTROL USING A SINGLE RESISTOR

Figure 4.14 shows a single resistor controlling the current that flows through a filament lamp. In electronic circuits there are many components that could replace the lamp. For example, a motor, a light emitting diode (LED), etc.

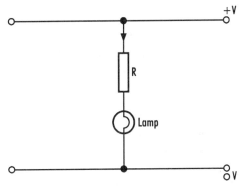

Fig. 4.14

Example

1. Consider the circuit diagram in Fig. 4.14 with a supply of 15 V. The lamp is rated at 6 V 25 mA. Calculate: (a) the potential difference across the resistor; (b) the value of the resistor that will limit the current flowing through the lamp to 25 mA.

(a) Supply $= 15 \text{ V}$; lamp voltage $= 6 \text{ V}$; current $I = 25 \text{ mA} = 0.025 \text{ A}$. Because it is a series circuit the potential difference across the resistor will be

$$V_R = 15 - 6 = 9 \text{ V}$$

(b) From Ohm's law:

$$R = \frac{V_R}{I} = \frac{9}{0.025} = 360 \Omega$$

Exercise 4.2

1. Describe, with the aid of sketches, the construction and action of a basic and a commercial potential divider.

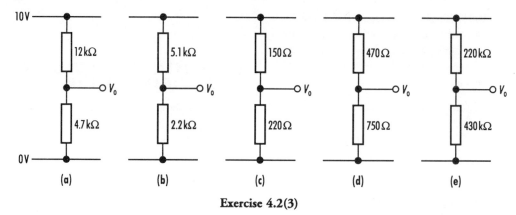

Exercise 4.2(3)

2. Draw a circuit diagram using two resistors connected as a potential divider. Identify the position of the input and output voltage.

3. The figure shows a number of potential dividers. For each case calculate the output voltage.

4. Draw a circuit diagram of a potential divider. The top resistor has a value of 33 kΩ and the bottom resistor has a value of 22 kΩ. If the output voltage is 10 V calculate the input voltage.

5. Explain, with the aid of a graphical symbol, the basic construction and action of a VR.

6. Describe, with the aid of sketches, the following: (a) slider-type VR; (b) lug-type wire-wound adjustable resistor; (c) wire-wound power rheostat; (d) enclosed VR; (e) skeleton-type pre-set trimmer.

7. Explain, with the aid of a sketch, how a VR can be used in a circuit to: (a) control current; (b) control voltage.

8. A VR of total resistance 18 kΩ is connected in series with 1 kΩ resistor. The input voltage is 15 V. If the VR is set in the ratio top:bottom as 3:1 draw a circuit diagram and determine the output voltage. What is the function of the 1 kΩ resistor?

9. A VR has a total resistance of 5.6 kΩ. It is connected in series with a resistor of value 1.2 kΩ. The input voltage is 12 V. Draw a circuit diagram and calculate the current flowing through the 1.2 kΩ resistor if the VR is set to: (a) its maximum value; (b) its minimum value.

10. Explain, with the aid of a circuit diagram, how a resistor can be used to control current when used in conjunction with another component.

11. A lamp rated at 6 V 50 mA is connected in series to a resistor of unknown value. A 12 V supply is connected across the series circuit. Draw a labelled circuit diagram. State the value of current flowing through the resistor. Calculate the value of the potential difference across the resistor and the resistance of the resistor.

12. Copy out and complete the following sentences.

Two resistors connected in series with a voltage supply make a ....
    The resistor connected to the positive supply rail is called the ...... resistor. The resistor connected to the negative supply rail is called the ...... resistor.

13. The figure shows a filament lamp and a bell connected to a 6 V supply. The filament lamp is rated at 6 V 50 mA and the bell at 6 V 25 mA. (a) should the ammeter have a high or low resistance? (b) are the lamp and bell connected in series or parallel? (c) state which switch must be closed to make the

filament lamp work; (d) state which switch must be closed to make the bell ring; (e) state the ammeter reading for the following switch conditions: (i) both switches closed; (ii) both switches open; (iii) switch 1 closed, switch 2 open; (iv) switch 1 open, switch 2 closed.

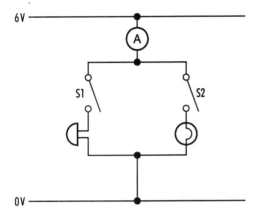

Answers

Exercise 4.1
1. See 4.1
2. See 4.1.1, 4.1.2, 4.1.3, 4.1.4
3. Power consumed 0.36 W, suitable
4. Power consumed 3 W, unsuitable
5. Suitable/suitable/suitable/unsuitable/ unsuitable/suitable
6. 470 Ω ± 5%
7. 510RK
8. 0.82 Ω, 2 Ω, 37 Ω, 160 Ω, 1.8 kΩ, 5.6 kΩ, 2.2 MΩ,   56 Ω ± 5%,   820 Ω ± 10%, 3.9 kΩ ± 5%,   8.2 kΩ ± 20%,   2.4 MΩ± 10%
9. R18, 1R0, 5R6, 20R, 220R, 1K5, 33K0, 5M6, 51RJ, 620RK, 750RM
10. (a) Brown, Green, Black, Gold; (b) Brown, Black, Brown, Silver; (c) Orange, White, Red; (d) Brown, Red, Orange, Gold; (e) Red, Red, Yellow, Silver; (f) Blue, Grey, Green
11. 10 Ω ± 20%, 100 Ω ± 20%, 470 Ω ± 20%, 20 kΩ ± 20%,   2.2 kΩ ± 20%,   3.3 kΩ ± 20%, 10 kΩ ± 20%, 470 kΩ ± 20%, 75 Ω ±

10%, 680 Ω ± 5%, 9.1 kΩ ± 5%, 39 Ω ± 10%
12. 12 Ω,   150 Ω,   180 Ω,   2200 Ω,   22 kΩ, 330 kΩ, 4.7 MΩ, 82 MΩ
13. Max. value 220 Ω, min. value 180 Ω
14. (a) 105 Ω, 95 Ω; (b) 132 Ω, 108 Ω; (c) 180 Ω, 120 Ω; (d) 1680 Ω, 1520 Ω; (e) 1980 Ω, 1620 Ω; (f) 24 kΩ, 16 kΩ; (g) 252 kΩ, 228 kΩ; (h) 330 kΩ, 270 kΩ; (i) 432 kΩ, 288 kΩ; (j) 4.515 MΩ, 4.085 MΩ; (k) 6.160 MΩ, 5.04 MΩ; (l) 7.44 MΩ, 4.96 MΩ; (m) 71.4 MΩ, 64.6 MΩ; (n) 82.5 MΩ, 67.5 MΩ; (o) 98.4 MΩ, 65.6 MΩ
15. 160RK, 144 Ω, 176 Ω
16.

Resistor value (Ω)	5% tolerance		10% tolerance		20% tolerance	
	max. (Ω)	min. (Ω)	max. (Ω)	min. (Ω)	max. (Ω)	min. (Ω)
10	10.5	9.5	11	9	12	8
120	126	114	132	108	144	96
1500	1575	1425	1650	1350	1800	1200
18	18.9	17.1	19.8	16.2	21.6	14.4
220	231	209	242	198	264	176
2700	2835	2565	2970	2430	3240	2160
33	34.65	31.35	36.3	29.7	39.6	26.4
390	409.5	370.5	429	351	468	312

Exercise 4.2
1. See 4.5.1 and 4.5.2 and Figs 4.5 and 4.6
2. See Fig. 4.5
3. (a) 2.814 V; (b) 3.014 V; (c) 5.946 V; (d) 6.148 V; (e) 6.615 V
4. See Fig. 4.5, 25 V
5. See Fig. 4.7
6. See 4.6.1, 4.6.2, 4.6.3, 4.6.4, 4.6.5
7. See 4.7.1 and 4.7.2
8. See 4.7.2 Example 1; 3.552 V
9. (a) 1.7647 mA; (b) 10 mA
10. See 4.8 and Fig. 4.14
11. See Fig. 4.14, 50 mA, 6 V, 120 Ω
12. Potential divider. Pull-up resistor and pull-down resistor.
13. (a) high; (b) parallel; (c) switch S2; (d) switch S1; (e) 75 mA, zero, 25 mA, 50 mA

5

COMPONENTS 2

5.1 ELECTROSTATICS

Fundamental to the study of capacitors and capacitance is electrostatics which deals with electric charges at rest. A negative charge is induced on an ebonite rod by rubbing it with fur. In this case electrons are transferred to the rod from the fur, resulting in an excess of electrons at its surface.

A positive charge is induced on a glass rod by rubbing it with silk. In this case electrons are removed from the rod by the silk resulting in a deficiency of electrons at its surface.

Bodies with the same type of charge, i.e. both positive or both negative, repel one another. Bodies with opposite charges, i.e. one positive and one negative, attract one another.

Figure 5.1 shows electric charges under differing circumstances with the arrows representing the direction of force on the charge.

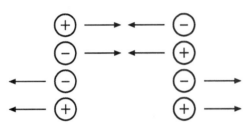

Fig. 5.1

5.2 CAPACITOR

A capacitor is a device that stores electric charge. In its simplest form a capacitor consists of two metal plates separated by insulating material which may be solid or liquid or just air. This separating material is known as the dielectric. A basic outline drawing of a parallel-plate capacitor is shown in Fig. 5.2.

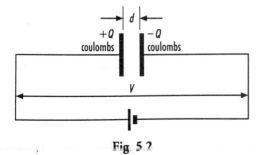

Fig. 5.2

5.3 CAPACITANCE

The property of a capacitor to store an electric charge with excess of electrons on one plate and a deficiency of electrons on the other is called its capacitance (C).

By experiment it can be shown that the quantity of charge (Q) is directly proportional to the applied voltage (V) or

$$Q \propto V$$

Therefore

$$Q = V \times \text{a constant} = CV$$

where C is the capacitance measured in farads (F).

A smaller unit of capacitance is the microfarad (μF) and a smaller unit still is the picofarad (pF).

$$1\,000\,000\,\mu F = 1\,F$$

$$1\,000\,000\,000\,000\,pF = 1\,F$$

The graph in Fig. 5.3 shows the relationship between the quantity of charge and applied voltage.

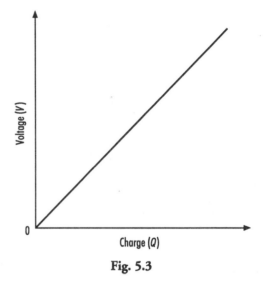

Fig. 5.3

Examples

1. A capacitor having a capacitance of 50 μF is connected across a 250 V d.c. supply.

Calculate the charge.

$$C = 50\,\mu F \quad V = 250\,V$$

Using $Q = CV$

$$Q = 50 \times 10^{-6} \times 250$$

$$= 12\,500 \times 10^{-6} = 0.0125\,C$$

2. A charge of $10\,\mu C$ causes the difference in potential between two plates to change by $2\,V$. Determine the capacitance.

$$Q = 10\,\mu C \quad V = 2\,V$$

From $Q = CV$

$$C = \frac{Q}{V} = \frac{10 \times 10^{-6}}{2}$$

$$= 5 \times 10^{-6}\,F = 5\,\mu F$$

Exercise 5.1

1. Calculate the charge on a $5\,\mu F$ capacitor when the applied voltage across the plates is $200\,V$.

2. A $5\,\mu F$ capacitor consists of two metal plates spaced $2\,mm$ apart, the dielectric being air. Calculate the applied voltage if the capacitor is charged to $10\,C$. What would be the electric field strength?

3. Calculate the capacitance of a capacitor charged to $0.03\,C$ by an applied potential difference of $1000\,V$.

4. (a) Explain the meaning of the term capacitance. (b) Describe one form of capacitor. (c) A capacitor has a capacitance of $50\,\mu F$. Calculate the charge it will receive when connected to a $200\,V$ d.c. supply.

5.4 ELECTROLYTIC CAPACITOR

An electrolytic capacitor is polarized which means that it must be connected into a circuit the correct way round. If, mistakenly, a voltage is applied to the capacitor in the reverse direction, the capacitor will be damaged and also other components in the circuit could be damaged. The damage concerned may extend even to the electrolytic capacitor exploding. All polarized capacitors are clearly marked with positive ($+$) and negative ($-$) terminals. The capacitor will have a maximum voltage rating with a working voltage which should never be exceeded.

The characteristic of the electrolytic capacitor is shown in the graphical symbol found in Fig. 1.5 by representing the positive ($+$) plate as an open rectangle. The electrolytic capacitor is a fixed capacitor in which the dielectric is a thin film of oxide deposited by electrolytic action upon aluminium foil which acts as the positive plate. The negative plate is a non-corrosive electrolyte. The electrolyte can be either a liquid or a paste, which saturates a piece of paper or gauze.

Because the dielectric is very thin this form of construction gives a high capacitance in a component of relatively small dimensions.

Aluminium electrolytic capacitors are manufactured in many different ways and provide a wide range of characteristics. A few types are considered.

5.4.1 Axial lead aluminium electrolytic

The general dimensions for this type of capacitor are shown in Fig. 5.4. A general technical specification would normally include

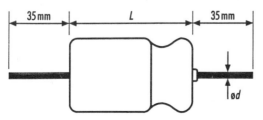

Fig. 5.4

the following:

Tolerance	±20%
Temperature range	−40 to +85°C
Shelf life	Minimum two years
d.c. working voltage range	6.3 V to 450 V
Capacitance range	100 μF to 4700 μF

Typical dimensions for a 100 μF 25 V capacitor are:

length $L = 13$ mm
diameter $D = 8$ mm
diameter $d = 0.6$ mm

This type of capacitor provides high general performance with reliability and is usually provided with tinned copper leads. Because of their high capacitance, electrolytic capacitors can be used in smoothing and decoupling circuits.

5.4.2 Radial electrolytic capacitors

The general dimensions for this type of capacitor are shown in Fig. 5.5. Typical dimensions for a miniature 100 μF 25 V capacitor are:

length $L = 11.5$ mm
diameter $D = 6$ mm
diameter $d = 2.5$ mm

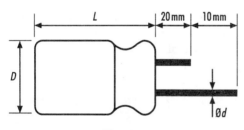

Fig. 5.5

The difference with the layout of the leads of the radial capacitor as compared to the axial lead capacitor is easily seen. With the radial capacitor both leads come out of the capacitor at the same end.

A general technical specification would normally include the following:

Tolerance	±20%
Temperature range	−40 to +85°C
Leakage current	4 μA

Some manufacturers do in fact make a high-temperature radial electrolytic capacitor with flame-retardant sleeves and case that has an operating temperature up to 105°C.

d.c. working voltage range	6.3 V to 100 V
Capacitance range	0.1 μF to 1200 μF

The sub-miniature type of electrolytic capacitor gives a high performance and is for use in applications where board space and height are at a premium.

5.4.3 Can-type electrolytic capacitor

As the name suggests this type of capacitor looks like the shape of a small tin can and is shown in Fig. 5.6.

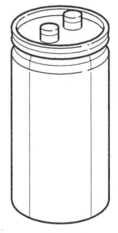

Fig. 5.6

One type of can capacitor has standard twist-prong mounting lugs with solder terminals using etched cathode construction for a hum-free operation and having a vent and seal design. Another type is used for printed circuit board (PCB) mounting, with long life, high capacitance values but small in physical size. A range of applications are available, such as in power supplies in digital equipment, switch-mode power supplies, storing energy in pulse systems and filters in measuring and control equipment. To ensure that the terminals are connected correctly the printed wire terminals are marked with a '1' for the positive (+) terminal and a '5' for the negative terminal.

Typical dimensions for a 100 μF 385 V capacitor are 25 mm diameter and 45 mm long with an operating temperature of −40 to +85°C. The d.c. leakage current is +4 μA and the operating life is 5000 hours.

5.4.4 Computer-grade electrolytic capacitor

This type of capacitor is used not only in computers but in many general applications. The outside appearance is very similar to the one shown in Fig. 5.6. The capacitor is designed to give a high performance and is usually sealed in a plastic sleeve. They generally have a maximum leakage current of 6 mA at 25°C with an expected very long life running into years.

Typical dimensions for a 100 μF 450 V capacitor are 35 mm diameter and 54 mm long.

One use of this type of capacitor is as an output capacitor in a high-frequency switching mode power supply.

5.4.5 Surface-mounted electrolytic capacitor

A general-purpose surface-mounted aluminium electrolytic chip capacitor is shown in outline in Fig. 5.7. Typical dimensions for a 10 μF 16 V capacitor are:

length $L = 12$ mm
width $W = 4$ mm
height $H = 4$ mm

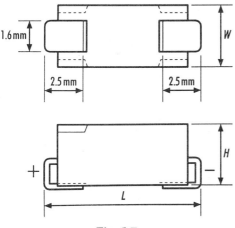

Fig. 5.7

The range of the operating temperature is −40 to +85°C with a wide tolerance range of −10% to +50%. It is important to remember that this is a polarized capacitor so care must be taken when making connection. The positive (+) terminal is marked together with the use of a bevelled edge to help with identification.

5.4.6 Tantalum capacitor

In this capacitor the metal electrode is made from tantalum instead of aluminium. Tantalum is a greyish-white metal that is very ductile and malleable. The electrolyte is either in liquid or paste form and is allowed to soak into a piece of paper or gauze. The tantalum capacitor can be supplied in several different formats, in a like manner to those previously discussed. A type not yet mentioned is a bead capacitor as shown in Fig. 5.8. Typical dimensions for a 100 μF 10 V capacitor are:

width $D = 8.5$ mm
height $H = 14$ mm

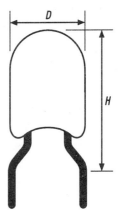

Fig. 5.8

A popular use for this radial-lead, solid tantalum capacitor is for use in entertainment and commercial equipment. They are flame-retardant, moisture-resistant and covered in an epoxy resin so that they do not chip or crack under high temperatures. The temperature range is from −55 to +125°C with a manufacturer's tolerance of ±20%.

Specific applications are in filter circuits, by-pass circuits, timing circuits, coupling and blocking circuits, etc.

5.5 NON-POLARIZED CAPACITORS

The non-polarized capacitor can be connected either way round in a circuit. The capacitors that we will now be considering will be marked with a colour code or have a value of capacitance marked on them. The working and maximum voltage will also be displayed. You are reminded that a capacitor consists of two plates separated by an insulating material called a dielectric.

The dielectric describes the type of capacitors in use. The dielectric is a material that can resist strong electric fields without breaking down. A measure of this strength is called the dielectric constant or relative permittivity. All materials have different values for their relative permittivity. The higher the relative permittivity the better is the dielectric.

A number of non-polarized capacitors are now considered.

5.5.1 Ceramic capacitor

This capacitor uses a ceramic as the dielectric material. Probably the most well-known ceramic is porcelain. Others are called steatite, forsterite, cordierite and alumina. A number of different types of ceramic capacitor are available.

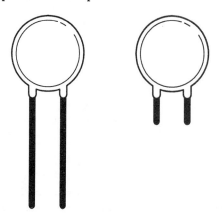

Fig. 5.9

(a) Ceramic disc

This type of ceramic capacitor is also known as a plate capacitor. As the name suggests it looks like a disc or plate but small in size. One feature of this capacitor is its high capacitance per unit volume which makes it suitable for use in high-density PCB applications. Figure 5.9 shows a typical long- and short-lead radial ceramic capacitor.

Depending on the performance required there is available a low-, medium- or high-permittivity capacitor. General details for each type of permittivity are now given:

Low relative permittivity

Working voltage	50 V d.c.
Operating temperature	−25 to +35°C
Insulation resistance	10 000 MΩ
Range	2 pF to 330 pF
Lead length	25 mm
Lead diameter	0.5 mm
Disc diameter	5–20 mm

This type of capacitor is often used when high stability, close tolerance and low losses are important in tuned circuits and filters or coupling and decoupling in high-frequency circuits.

Medium relative permittivity   If we consider the same working voltage of 50 V d.c. then the capacitor range increases up to 4700 pF with the same value of insulation resistance but with an operating temperature going up to +85°C. In this case the tolerance is ±10%. This type of capacitor is suitable for applications where a small non-linear change of capacitance with temperature is acceptable and very low losses are not essential to the satisfactory performance in a circuit. The physical dimensions are much the same as those quoted in the previous case.

High relative permittivity   We will again consider the same working voltage of 50 V d.c. The capacitance range goes up to 100 000 pF with the same insulation resistance and operating temperature as the medium relative permittivity capacitor but with a tolerance of −20% to +80%. This type of capacitor is best used where low losses and high stability are not critical to the circuit function. The capacitor has an external protective coating and is high-temperature wax impregnated.

There is available a high-voltage range of ceramic disc capacitors going up to 15 000 V. A typical 15 000 V capacitor would have a capacitance of up to 1000 pF with an operating temperature up to +85°C. The quoted tolerance is similar to that of the high relative permittivity type mentioned previously. Because this type of capacitor often has low inductance it is suitable for high-frequency decoupling and suppression in logic circuits.

(b) Multi-layer

A multi-layer ceramic chip surface-mounted capacitor is shown in outline in Fig. 5.10. Typical dimensions for a 50 V capacitor of tolerance ±10% with a temperature range of −55 to +125°C are:

length $L = 2$ mm
width $W = 1.5$ mm
thickness $T = 1.3$ mm

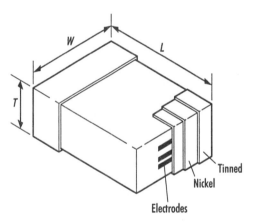

Fig. 5.10

This type of capacitor offers a range of working voltages and is very suitable when high performance and excellent reliability are needed.

5.5.2 Plastic film capacitor

This is a capacitor that uses a plastic film as its dielectric. The dielectric materials in use are polycarbonate, polyester, polypropylene and polystyrene. No emphasis has been placed on order of importance, they are simply listed in alphabetical order. Each type will now be considered.

(a) Polycarbonate

Several shapes of this type of capacitor are available.

Axial   The physical dimensions are 4–15 mm diameter with a length of 10–34 mm. They are relatively small, light-weight and have high capacitance values from 1000 pF to 10 000 pF. They tend to be stable with a long life and have a low dissipation factor with an operating temperature range of −55 to +125°C. A typical working voltage is 50 V d.c. with a capacitance tolerance of ±5%.

Radial   A radial-type capacitor is shown in outline in Fig. 5.11. Typical dimensions for a 10 nF capacitor $(1 \text{ nF} = 10^{-9} \text{ F})$ are width 7 mm, height 10 mm and thickness 5 mm. The lead length is 6 mm with a diameter of 0.5 mm.

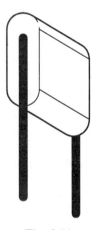

Fig. 5.11

The capacitor electrodes are metal foil with a polycarbonate dielectric. These capacitors have better electrical characteristics than polyester ones with excellent stability at high frequencies. For a working voltage of 100 V you can expect to use them in a temperature range of −55 to +100°C with a capacitance tolerance of ±10%. They have an external flame-retardant case with an epoxy resin seal. They are very useful in filter, timing and other high-stability applications.

Motor-run can capacitors   This capacitor is a stud mounting film capacitor suitable for a wide range of motor applications. It is shaped like a tin can with typical dimensions for a 280 V a.c. capacitor of length 40 mm and diameter 30 mm. The capacitors are provided with solder tab terminations and are used in a.c. applications such as filters, phase shifting or power factor

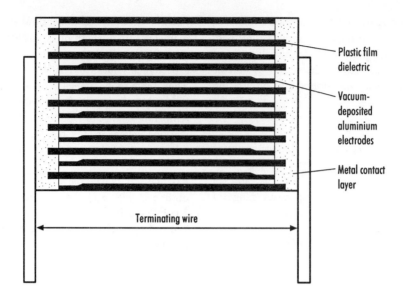

Plastic film dielectric

Vacuum-deposited aluminium electrodes

Metal contact layer

Terminating wire

Fig. 5.12

correction, and d.c. or pulse applications such as thyristor commutation.

(b) Polyester

Two basic shapes of this capacitor are available.

Axial   The physical dimensions are 3–8 mm diameter with a length of 8–30 mm. A typical capacitance range is from 2.2 nF to 220 nF. They have high-frequency performance and are wound with very thin polyester film dielectric and thin gauge foil but have a rugged outer sleeve with an epoxy end-fill. By nature of the dimensions they are suitable for automated installation on printed wiring boards. The normal operating temperature is between −55 and +85°C.

Radial   The shape is similar to that shown in Fig. 5.11. A wide range of voltages are available. For a 10 nF capacitor rated at 400 V d.c. typical dimensions are width 12 mm, height 12 mm and thickness 4 mm with a lead diameter of 0.6 mm. The insulation resistance is very high and a variety of capacitance tolerance ranges are available. The polyester capacitor gives high performance and is suitable for general purpose applications such as coupling and decoupling.

Miniature versions are available where weight

and volume are critical. Most types are encapsulated in a vacuum cast flame-retardant case and they are resistant to modern cleansing solvents. Crimped and cropped lead types are available for mounting on PCBs. Stability and quality are usually very good.

A typical internal structure is shown in Fig. 5.12. The terminating wire makes electrical contact with the complete end surface of the electrode which reduces inductance to a minimum.

(c) Polypropylene

Two basic shapes of this capacitor are available.

Axial   These capacitors are coated with epoxy lacquer making them water-repellent, solvent and acid-resistant and usually having good resistance to thermal shocks. They have 30 mm long leads, solder-coated, which makes them suitable for horizontal or vertical mounting on PCBs. The typical dimensions for a 150 pF 400 V d.c. capacitor are length 12 mm and diameter 5 mm. For values less than 100 pF they have a tolerance of ±1%. The insulation resistance is high, in the region of 100 000 MΩ with the normal operating temperature range of −40 to +100°C. The outer case is flame-retardant and has epoxy resin end seals.

Radial The shape is similar to that shown in Fig. 5.11. A wide range of voltages are available from 63 V d.c. to 2000 V d.c. For a 10 nF capacitor rated at 63 V d.c. typical dimensions are width 7 mm, height 8 mm and thickness 6 mm with a lead length of 6 mm and diameter 0.5 mm.

The capacitor electrodes are metal foil with a polypropylene dielectric. The dielectric is strong and has low loss with a high insulation resistance in the area of 500 000 MΩ. With a 100 nF capacitor working at 1000 V d.c. you can expect a tolerance of ±5%. The operating temperature is in the range −55 to +85°C. The polypropylene capacitor gives a very good high-frequency performance and will withstand high voltages and is used in tuned circuits, filter networks, timing networks, oscillator circuits and other high-performance applications.

Externally the capacitor is encapsulated in a flame-retardant plastic case with an epoxy resin end seal.

(d) Polystyrene

This is probably the most well known dielectric material because it is used in so many everyday situations from packing in boxes to ceiling tiles. Two basic shapes of this capacitor are available.

Axial They are manufactured from a stretched, fused polystyrene that provides temperature stability and humidity protection. The dielectric has low loss and low absorption, good long-term stability with a very high insulation resistance in the area of 500 000 MΩ. A 100 pF capacitor working at 160 V d.c. is typically 8 mm long and 4 mm diameter. A typical operating temperature is −40 to +85°C. This type of capacitor is used in tuned circuits, filter networks, timing circuits, etc. It has a relatively low cost and is physically smaller than an equivalent mica capacitor.

Radial The shape is similar to that of Fig. 5.11. This shaped capacitor has the same overall characteristics and uses as the axial type of capacitor. In addition this type is suitable for use in telephone equipment. The physical dimensions of a 100 pF 63 V d.c. capacitor are width 8 mm, height 8 mm and thickness 5 mm.

5.5.3 Air capacitor

As the name suggests, this is a capacitor that uses air as its dielectric material. Figure 5.13 shows the outline of a variable capacitor using air as the dielectric.

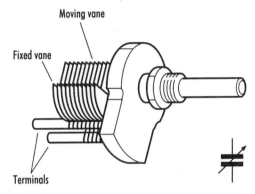

Fig. 5.13

The principal use of a variable capacitor is as a tuning device, often used in radios. The capacitance varies with the area of the plates facing each other. When the moving vane and the fixed vane are fully meshed together the capacitance will then be at its maximum. If the moving vane is slowly rotated so that the vanes are not fully meshed the capacitance will gradually decrease. The British Standard graphical symbol is shown by the side of the variable capacitor in Fig. 5.13.

5.5.4 Trimming capacitor

This capacitor is a fairly small variable capacitor used in parallel with a large fixed capacitor. The trimming capacitor is adjusted so that the total capacitance of the circuit can be set with a reasonable accuracy. Trimming capacitors are manufactured using air, ceramic or polyethylene as the dielectric material.

5.5.5 Mica capacitor

This is a capacitor that uses mica as the dielectric material. Mica itself is a very complex aluminosilicate of potassium, magnesium and iron but for use as a dielectric it is quite straightforward being produced in sheets. In general terms mica capacitors have characteristics of low loss, good frequency stability and have a coating that conforms to most humidity requirements. Their normal temperature operating range is −55 to +125°C. One manufacturer will provide mica capacitors in the tolerance range D, F, G, J. If you do not remember what this means then go back to Section 4.2 on the British Standard code.

The construction of a standard mica capacitor consists of sheets of mica clamped between tin foil electrodes. Other types of construction are available such as eyelet-construction mica capacitors, bonded-silvered mica capacitors, and button mica capacitors. Information on these types of capacitors can be found in specialist capacitor booklets.

5.5.6 Paper capacitors

As you would probably expect this capacitor uses paper as its dielectric material. The capacitor is made by winding together aluminium foil with a layer of paper. Paper is very vulnerable to moisture so to overcome this problem the paper is impregnated with a suitable wax or oil. In general terms this type of capacitor has medium losses and medium capacitance stability. The insulation resistance does fall rapidly if exposed to high temperatures. It can be used in high a.c. and d.c. voltage applications. The capacitor is relatively cheap and can be used as a general-purpose capacitor in power factor correction, blocking and frequency by-pass circuits.

5.6 CAPACITOR COLOUR CODE

Most capacitors have the capacitance value marked on them because there is no agreed colour code. However, some capacitors do use a colour code using the same system as the resistor colour code. The colours represent the numerical values measured in picofarads. For convenience the colour code is repeated. The colours used to represent the different numbers are given in Table 5.1.

Table 5.1

Black	0	Violet	7
Brown	1	Grey	8
Red	2	White	9
Orange	3	Gold	±5%
Yellow	4	Silver	±10%
Green	5	No colour	±20%
Blue	6		

If, A, B, C and D are the coloured bands, then A and B represent the first two numbers of the capacitance. Band C is the number of zeros and the fourth band is the tolerance of the capacitor.

Examples

1. Write down the value of a capacitor colour coded: Brown, Black, Orange, Gold.

    Brown indicates   1
    Black indicates   0
    Orange indicates   3 noughts
    Gold indicates   5%

    Capacitor value   10 000 pF ± 5%

2. Write down the colour code for a capacitor of value 4700 pF ± 10%.

    Number 4 indicates   Yellow
    Number 7 indicates   Violet
    Two noughts indicates   Red
    ±10% indicates   Silver

    Colour code is Yellow, Violet, Red, Silver.

3. Give the value of the capacitor in picofarads and microfarads colour coded Blue, Grey, Yellow.

    Blue indicates   6
    Grey indicates   8
    Yellow indicates   4 noughts
    No colour indicates ±20%

    Capacitor value   680 000 pF ± 20%   or 0.68 μF ± 20%

    Another colour code that is in use for polyester capacitors uses five bands. Bands A and B use the colours already listed. Band C is:

Orange   ×0.001 μF
Yellow   ×0.01 μF
Green   ×0.1 μF

Band D is tolerance:

White   ±10%
Black   ±20%

Band E is working voltage:

Red   250 V d.c.
Yellow   400 V d.c.

Examples

1. Write down the colour code for a 7 μF capacitor with a tolerance of ±10% and a working voltage of 250 V.

Violet indicates   7
Black indicates   0
Green indicates   ×0.1 μF
White indicates   ±10%
Red indicates   250 V d.c.

Colour code is Violet, Black, Green, White and Red.

2. Write down the value of a capacitor colour coded: Blue, Grey, Orange, Black, Yellow.

Blue indicates   6
Grey indicates   8
Orange indicates   ×0.001 μF
Black indicates   ±20%
Yellow indicates   400 V d.c.

Capacitor value   0.068 μF ± 20%   400 V d.c.

Exercise 5.2

1. Explain what is meant by the term electrostatics.

2. Explain, with the aid of a sketch, the basic action and construction of a capacitor.

3. Explain the difference between an electrolytic capacitor and a non-electrolytic capacitor.

4. Draw the British Standard graphical symbols for: (a) a capacitor general symbol; (b) a polarized capacitor general symbol; (c) a polarized electrolytic symbol.

5. Explain, with the aid of a sketch, the construction and action of an axial lead and radial lead aluminium electrolytic capacitor. Write out a general technical specification for such a capacitor.

6. Explain, with the aid of a sketch, the construction and action of the following types of capacitor: (a) can-type electrolytic; (b) computer-grade electrolytic; (c) surface-mounted electrolytic; (d) tantalum electrolytic.

7. State the purpose of a dielectric material in a capacitor.

8. Explain, with the aid of a sketch, what is meant by the term **ceramic disc capacitor**.

9. State the difference between a low-, medium- and high-permittivity ceramic capacitor.

10. Sketch and describe a multi-layer ceramic chip capacitor.

11. What is meant by the term plastic film capacitor?

12. Explain, with the aid of a sketch, the following types of plastic film capacitors: (a) polycarbonate; (b) polyester; (c) polypropylene; (d) polystyrene.

13. What is an air capacitor?

14. Explain how a trimming capacitor is used.

15. Describe the constructional details of a mica and a paper capacitor.

16. Write down the colour code that should be used for the following capacitors: (a) 100 pF; (b) 1200 pF; (c) 15 000 pF; (d) 180 000 pF; (e) 2 200 000 pF; (f) 27 000 000 pF; (g) 33 pF; (h) 390 pF; (i) 4700 pF; (j) 56 000 pF; (k) 680 000 pF; (l) 8 200 000 pF.

17. What tolerance does band four signify when it is coloured (a) Gold and (b) Silver?

18. Write down the colour code that would be used for the following capacitors: (a) 160 pF ± 5%; (b) 2000 pF ± 5%; (c) 360 000 pF ± 5%; (d) 7 500 000 pF ± 10%; (e) 91 000 000 pF ± 10%; (f) 110 pF; (g) 1300 pF; (h) 24 000 pF.

19. State the values in picofarads and microfarads for capacitors with the following colours: (a) Brown, Black, Orange; (b) Red, Red, Orange; (c) Brown, Green, Yellow; (d) Brown, Green, Blue, Gold; (e) Orange, White, Brown, Gold, (f) Grey, Red, Yellow, Silver; (g) Green, Blue, Black, Silver.

5.7 CAPACITOR FIELD STRENGTH

The field strength (E), sometimes known as electric force, is defined as the potential difference per unit thickness in a dielectric. It is given by

$$E = \frac{\text{potential difference } (V)}{\text{distance between plates } (d)}$$

or

$$E = \frac{V}{d} \text{ volts/metre}$$

Example

1. A capacitor has two plates separated by a dielectric 2.5 mm thick. If the potential difference applied between the plates is 2000 V, calculate the electric field strength in the dielectric.

$$V = 2000 \text{ V} \quad d = 2.5 \times 10^{-3} \text{ m}$$

Using $E = V/D$

$$E = \frac{2000}{2.5 \times 10^{-3}} = \frac{2000 \times 10^3}{2.5}$$

$$= 800 \times 10^3 \text{ V/m} = 800 \text{ kV/m}$$

Exercise 5.3

1. Calculate the electric field strength in a dielectric that is 5 mm thick when a potential difference of 250 V is applied across it.

2. A capacitor has two plates separated by a dielectric 4 mm thick. If the potential difference applied between the plates is 1000 V, determine the electric field strength.

3. Two metal electrodes separated by a distance of 1 mm are connected to a 200 V supply. For the electric field existing between the electrodes calculate the electric field strength.

4. A potential difference of 2 kV exists between two metal plates in an electric field situated 8 mm apart. Calculate the magnitude of the electric field strength.

5. A capacitor is supplied at 300 V and has a dielectric of 0.15 mm in thickness. Calculate the stress in the dielectric in kV/mm.

6. (a) Explain how, by a simple demonstration, it can be shown that energy is stored in a capacitor. (b) If the thickness of the dielectric material in a capacitor was 0.2 mm $(0.2 \times 10^{-3} \text{ m})$ and a potential difference of 1000 V was applied across the dielectric, what would be the electric field strength or potential gradient in volts per metre?

7. Why is it essential that the dielectric stress must not exceed a certain value for a given insulating material?

5.8 ELECTRIC FLUX AND ELECTRIC FLUX DENSITY

From Fig. 5.2 it can be seen that the charge on both plates is Q coulombs. An electric flux of Q coulombs is created by the charge Q coulombs. Let the area of one side of one plate, in square metres, equal a. Then electric flux density (D) is given by

$$D = \frac{Q}{a} \text{ coulombs/square metre}$$

Example

1. Two plates of a capacitor, each having an area of $0.05 \, \text{m}^2$, are charged so that each plate holds a charge of $15 \, \mu\text{C}$. Calculate the electric flux density in the dielectric separating the plates.

$$Q = 15 \times 10^{-6} \, \text{C} \quad a = 0.05 \, \text{m}^2$$

Using $D = Q/a$

$$D = \frac{15 \times 10^{-6}}{0.05} = 300 \times 10^{-6} \, \text{C/m}^2$$

$$= 300 \, \mu\text{C/m}^2$$

Exercise 5.4

1. A $2 \, \mu\text{F}$ capacitor charged by $500 \, \text{V}$ consists of two metal plates, each $0.05 \, \text{m}^2$ in area and separated $4 \, \text{mm}$ apart. Calculate: (a) quantity of charge; (b) electric field strength; (c) electric flux density.

2. A capacitor consists of two metal plates, each $200 \, \text{mm}$ by $150 \, \text{mm}$, placed parallel and $3 \, \text{mm}$ apart. It is now charged to a potential of $25 \, \text{V}$ and discharged through a galvanometer with a scale of $0.2 \, \mu\text{C}/$ division. If the galvanometer deflects by 80 divisions, what is the value of the capacitance? Calculate also the electric field strength and the electric flux density.

3. Explain the terms electric force, electric flux density, and permittivity.
   A $0.05 \, \mu\text{F}$ parallel-plate capacitor has a potential difference of $120 \, \text{V}$ applied across it. If the effective plate area is $0.06 \, \text{m}^2$, find: (a) the electric flux density; (b) the electric force between the plates, which are separated $0.05 \, \text{mm}$ apart.

5.9 RELATIONSHIP BETWEEN ELECTRIC FLUX DENSITY AND ELECTRIC FIELD STRENGTH

The ratio of electric flux density to electric field strength in free space is called the permittivity of free space (ϵ_0).

$$\epsilon_0 = \frac{\text{Electric flux density } (D)}{\text{Electric field strength } (E)} = \frac{Q/a}{V/d}$$

$$= \frac{dQ}{Va}$$

We can replace Q/V by C, then

$$\epsilon_0 = \frac{Cd}{a}$$

and capacitance is given by

$$C = \frac{a\epsilon_0}{d}$$

It can be shown by experiment that $\epsilon_0 = 8.854 \times 10^{-12} \, \text{F/m}$. The absolute permittivity $(\epsilon) = \epsilon_0 \epsilon_r$ where ϵ_r is the relative permittivity, often called the dielectric constant. Typical values are given in Table 5.2.

Table 5.2

Material	Relative permittivity (ϵ_r)
Air	1
Bakelite	5
Glass	7.5
Oil	3
Mica	5
Paper	2.5
Polythene	2.2
Porcelain	6.5
Rubber	2.75
Ebonite	2.8
Pure water	80

For any capacitor

$$C = \frac{\epsilon_0 \epsilon_r a}{d}$$

Examples

1. Calculate the capacitance of two metal plates each of area $40 \, \text{m}^2$ and separated by an air dielectric $4 \, \text{mm}$ thick.

$$a = 40\,\text{m}^2 \quad d = 4 \times 10^{-3}\,\text{m}$$

$$e_r = 1$$

Using $C = \epsilon_0 \epsilon_r a / d$

$$C = \frac{8.854 \times 10^{-12} \times 1 \times 40}{4 \times 10^{-3}}$$

$$= 88.54 \times 10^{-9}\,\text{F} = 88.54\,\text{nF}$$

2. A capacitor consists of two metal plates, each having an area of $0.2\,\text{m}^2$, spaced 3 mm apart. The whole of the space between the plates is occupied by paper having a relative permittivity of 2.5. If a potential difference of 500 V is maintained between the two plates calculate: (a) the capacitance; (b) the charge; (c) the electric field strength; (d) the electric flux density.

$$a = 0.2\,\text{m}^2 \quad d = 3 \times 10^{-3}\,\text{m} \quad \epsilon_r = 2.5$$

$$V = 500\,\text{V}$$

(a) Using $C = (\epsilon_0 \epsilon_r a)/d$

$$C = \frac{8.854 \times 10^{-12} \times 2.5 \times 0.2}{3 \times 10^{-3}}$$

$$= 1.476 \times 10^{-9}\,\text{F}$$

(b) Using $Q = CV$

$$Q = 1.476 \times 10^{-9} \times 500$$

$$= 738 \times 10^{-9}\,\text{C} = 0.738\,\mu\text{C}$$

(c) Using $E = V/d$

$$E = \frac{500}{3 \times 10^{-3}} = \frac{500 \times 10^3}{3}$$

$$= 166.6 \times 10^3\,\text{V/m} = 166.6\,\text{kV/m}$$

(d) Using $D = Q/a$

$$D = \frac{0.738}{0.2} = 3.69\,\mu\text{C/m}^2$$

Exercise 5.5

1. A capacitor has two plates each of area $0.005\,\text{m}^2$ and an air space between them of 4 mm. A constant voltage of 240 V is

applied across the plates. Calculate: (a) the capacitance in F, μF, pF and nF; (b) the charge in μC; (c) the electric flux density in μC/m^2; (d) the electric field strength in V/m.

2. A capacitor consists of two metal plates, each 150 mm × 150 mm, separated by an air dielectric. Calculate the distance between the plates if the capacitance is 400 pF.

3. A $2\,\mu$F capacitor is made from two strips of metal foil separated by a dielectric of permittivity 3. The dielectric material is 0.04 mm thick. If the width of each metal foil strip is 120 mm calculate the length of metal foil required to manufacture the capacitor.

4. A capacitor consists of two metal foil strips separated from each other by a waxed impregnated paper 0.12 mm thick having a dielectric constant of 2.4. If the capacitance of the capacitor is to be $3\,\mu$F, calculate the length of the paper dielectric if it is 120 mm wide.

5. A parallel-plate air-spaced capacitor is charged from a 200 V battery, which is then removed and the capacitor is lowered into a bath of oil without any leakage of electric charge in the process.
    What change occurs in: (a) the potential difference across the plates; (b) the electric flux density between the plates?
    Take the effective area of each of the two plates as $0.02\,\text{m}^2$ and their distance apart as 1 mm. The relative permittivity of the oil is 5.0.

6. Define the terms electric potential gradient and relative permittivity.
    A parallel-plate capacitor has an effective plate area of $0.1\,\text{m}^2$ (each plate), separated by a dielectric 0.5 mm thick. Its capacitance is 442 pF and it is charged to a potential difference of 10 kV. Calculate, from first principles: (a) the potential

gradient in the dielectric; (b) the electric flux density in the dielectric; (c) the relative permittivity of the dielectric material.

7. A capacitor is composed of two flat parallel plates separated by a slab of insulation. Explain why a current will flow for a short time when this capacitor is connected across a d.c. supply.

   In such a capacitor each of the plates is a rectangle 100 mm by 50 mm, the thickness of the insulation is 0.5 mm, and the total capacitance is 0.0005 μF. The capacitor is connected to a 100 V d.c. supply for some time. Calculate the dielectric flux density and the electric stress in the insulation, clearly stating the units in which each is measured.

   What would be the capacitance of this arrangement if the thickness of the insulation were to be increased to 1.0 mm?

8. Two capacitors each consist of a pair of parallel, flat metal plates each 150 mm^2. In one capacitor the plates are 2 mm apart in the air; in the other, the plates are 5 mm apart, the space between them being filled with a sheet of insulating material having a relative permittivity of 6. Calculate the ratio of the capacitances of these two capacitors and the ratio of the energies stored in them when a potential of V volts is applied to each.

9. A potential difference of 10 kV is applied to the terminals of a capacitor consisting of two circular plates, each having an area of 0.01 m^2, separated by a dielectric 1 mm thick. If the capacitance is $3 \times 10^{-4} \mu$F, calculate: (a) the total electric flux in coulombs; (b) the electric flux density; (c) the relative permittivity of the dielectric.

10. An electrostatic device consists of two parallel conducting plates each of area 0.1 m^2. When the plates are 10 mm apart in air the attractive force between them is 0.1 N. Calculate the potential difference between the plates. Find also the energy stored in the system.

If the device is used in a container filled with a gas of relative permittivity 4, what effect does this have on the force between the plates?

5.10 MULTI-PLATE CAPACITOR

A capacitor having n plates with adjacent plates separated by a distance d metres and alternate plates joined together as shown in Fig. 5.14 will be the same as $(n-1)$ capacitors connected in parallel. Therefore its capacitance will be given by

$$C = \frac{a\epsilon_0\epsilon_r(n-1)}{d}$$

Fig. 5.14

Example

1. Calculate the capacitance of a capacitor having seven parallel plates separated by a dielectric 0.2 mm thick. The area of one side of each plate is 0.004 m^2 and the relative permittivity of the dielectric is 5.

   $d = 0.2 \times 10^{-3}$ m $a = 0.004$ m^2 $\epsilon_r = 5$

   $n = 7$

   Using $C = [a\epsilon_0\epsilon_r(n-1)]/d$

   $$C = \frac{0.004 \times 8.854 \times 10^{-12} \times 5 \times (7-1)}{0.2 \times 10^{-3}}$$

   $$= 5.31 \times 10^{-9}\,\text{F} = 5.31\,\text{nF}$$

Exercise 5.6

1. A multi-plate capacitor has 11 plates, each pair of which is separated by a dielectric of thickness 0.4 mm and relative permittivity

2.2. The area of each plate side is 0.05 m². Calculate: (a) the capacitance of the capacitor; (b) the charge for an applied voltage of 400 V; (c) the electric field strength; (d) the electric flux density.

2. What are the factors which determine the capacitance of a parallel-plate capacitor? Mention how a variation in each of these factors will influence the value of capacitance.

Calculate the capacitance in picofarads of a capacitor having 11 parallel plates separated by mica sheets 0.2 mm thick. The area of one side of each plate is 0.001 m² and the relative permittivity of mica is 5.

5.11 ENERGY STORED IN A CAPACITOR

When a capacitor is being charged by an applied voltage the quantity of charge $Q = CV$ and at the same time $Q = It$.

Work done in charging the capacitor is stored in the capacitor as potential energy and released when the capacitor is discharged. A graph representing this situation is given in Fig. 5.15.

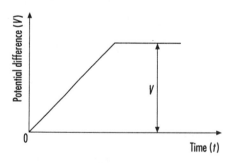

Fig. 5.15

Average potential difference across capacitor during charging $= V/2$ and average power to the capacitor during charging $= (V/2) \times I$. Total energy stored in the capacitor $(W) =$ power × time. Therefore

$$W = \frac{VIt}{2}$$

We can replace It by Q so

$$W = \tfrac{1}{2}QV$$

But $Q = CV$ so

$$W = \tfrac{1}{2}CVV = \tfrac{1}{2}CV^2$$

i.e. electrostatic energy stored in the capacitor $W = \tfrac{1}{2}CV^2$ joules.

Examples

1. Calculate the energy stored in a $2\,\mu F$ capacitor when it is connected across a 500 V d.c. supply.

$$C = 2 \times 10^{-6}\,\text{F} \quad V = 500\,\text{V}$$

Using $W = \tfrac{1}{2}CV^2$

$$W = \tfrac{1}{2} \times 2 \times 10^{-6} \times 500^2$$

$$= 500^2 \times 10^{-6} = 0.25\,\text{J}$$

2. A $10\,\mu F$ capacitor connected to a d.c. supply is charged with 0.005 C. Calculate: (a) the supply voltage; (b) the energy stored.

$$C = 10 \times 10^{-6}\,\text{F} \quad Q = 0.005\,\text{C}$$

(a) Using $Q = CV$

$$V = \frac{Q}{C} = \frac{0.005}{10 \times 10^{-6}}$$

$$= \frac{0.005 \times 10^6}{10} = 500\,\text{V}$$

(b) From $W = \tfrac{1}{2}CV^2$

$$W = \tfrac{1}{2} \times 10 \times 10^{-6}\ 500^2 = 1.25\,\text{J}$$

Exercise 5.7

1. Calculate the energy stored in a $5\,\mu F$ capacitor when it is connected across a 250 V d.c. supply.

2. A charge of $8\,\mu C$ causes the difference in potential between two conductors to change by 4 V. Calculate the capacitance and energy stored between the two conductors.

3. A 100 V d.c. supply is connected to an air-dielectric capacitor, consisting of two plates each of area $0.02\,\text{m}^2$. Calculate: (a) the capacitance; (b) the energy stored; (c) the electric field strength; (d) the electric flux density.

   If the air is now replaced by paper with permittivity 2.5 calculate the new values of (a) capacitance, (b) stored energy, (c) electric flux density.

4. An electrostatic device consists of two parallel conducting plates each of area $0.1\,\text{m}^2$. When the plates are 10 mm apart in air the attractive force between them is 0.15 N. Calculate the potential difference between the plates. Find also the energy stored in the system.

5.12 PURE CAPACITANCE ON A.C.

The circuit diagram is shown in Fig. 5.16.

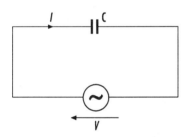

Fig. 5.16

When an a.c. circuit is purely capacitive the current (I) leads the voltage (V) by 90° or $\pi/2$ radians as shown in Fig. 5.17.

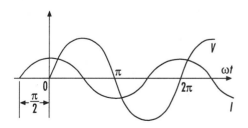

Fig. 5.17

The phasor diagram for this situation is shown in Fig. 5.18. A phasor is a line of length

proportional to the value of the alternating current or voltage it represents, rotating anti-clockwise about one of its ends at the frequency of the alternating current or voltage. It is usually drawn at the angle of phase corresponding to zero time.

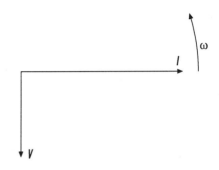

Fig. 5.18

A capacitor also offers opposition to an a.c. This opposition is termed capacitive reactance (X_c), given by

$$X_c = \frac{V}{I}$$

Capacitive reactance is inversely proportional to the product of capacitance and frequency, i.e.

$$X_c = \frac{1}{2\pi f C} = \frac{1}{\omega C}$$

The graphs shown in Figs 5.19 and 5.20 show the effect of frequency on capacitive reactance and current.

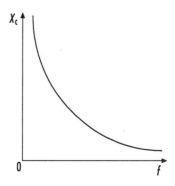

Fig. 5.19

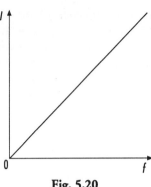

Fig. 5.20

From $X_c = 1/(2\pi f C)$

$$C = \frac{1}{2\pi f X_c}$$

$$= \frac{1}{2 \times 3.142 \times 50 \times 60}$$

$$= 0.000\,053\,\text{F}$$

$$= 53\,\mu\text{F}$$

Examples

1. Calculate the capacitive reactance of a capacitor taking a current of 2 A when connected to a 240 V 50 Hz supply.

$$V = 240\,\text{V} \quad I = 2\,\text{A}$$

Using $X_c = V/I$

$$X_c = \frac{240}{2} = 120\,\Omega$$

2. Determine the reactance of a $100\,\mu\text{F}$ capacitor when connected to a 240 V 50 Hz supply.

$$V = 240\,\text{V} \quad f = 50\,\text{Hz}$$

$$C = 100\,\mu\text{F} = 100 \times 10^{-6}\,\text{F}$$

Using $X_c = 1/(2\pi f C)$

$$X_c = \frac{1}{2 \times 3.142 \times 50 \times 100 \times 10^{-6}}$$

$$= \frac{10^6}{2 \times 3.142 \times 50 \times 100}$$

$$= 31.82\,\Omega$$

3. Calculate the capacitance of a capacitor taking a current of 4 A when connected to a 240 V 50 Hz supply.

$$V = 240\,\text{V} \quad I = 4\,\text{A} \quad f = 50\,\text{Hz}$$

Using $X_c = V/I$

$$X_c = \frac{240}{4} = 60\,\Omega$$

Exercise 5.8

1. Calculate the reactance of the following capacitors when connected to a 50 Hz supply: (a) 50 F; (b) 100 F; (c) $25\,\mu\text{F}$; (d) $75\,\mu\text{F}$.

2. Determine the capacitance of the following capacitors which have the following reactances at 50 Hz: (a) $250\,\Omega$; (b) $120\,\Omega$; (c) $35\,\Omega$; (d) $10\,\Omega$.

3. Calculate the value of a capacitor which will take a current of 20 A from a 240 V 50 Hz supply.

4. Define the term 'capacitive reactance'.

5. Calculate the voltage drop across a $4\,\mu\text{F}$ capacitor when a current of 0.5 A at 50 Hz flows through it.

6. Determine the value of a capacitor on a 100 V 50 Hz supply when the current flowing is 0.3 A.

7. A capacitor takes a current of 5 A from a 415 V 50 Hz supply. Determine the current taken if the supply falls to 400 V 50 Hz.

8. Determine the current taken by a $25\,\mu\text{F}$ capacitor when it is connected across a 240 V 50 Hz supply.

9. A $10\,\mu\text{F}$ capacitor is connected to a variable frequency 150 V supply. Draw a graph to show the variation in current through the capacitor as the frequency changes from 20 Hz to 60 Hz.

10. (a) A coil having a resistance of 75 Ω takes a current of 2 A when connected to a 240 V 50 Hz supply. Calculate the inductance of the coil. (b) A capacitor takes a current of 1.25 A when connected to a 250 V 50 Hz supply. Calculate the capacitance of the capacitor.

5.13 PURE CAPACITANCE ON D.C.

When a capacitor is connected to a d.c. supply, current flows, and the capacitor is fully charged when the current stops flowing. The capacitor can be discharged by placing, very carefully, a resistor across its terminals. Graphs of charge and discharge can be drawn as shown in the following sections.

5.14 THE SERIES *RC* CIRCUIT – CHARGING

When the switch is closed in position 'a' as shown in the circuit diagram of Fig. 5.21 the capacitor is being charged.

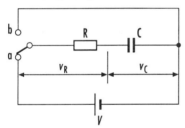

Fig. 5.21

The shape of the graph obtained when the potential difference of the capacitor is plotted against time during this charging period is shown in Fig. 5.22.

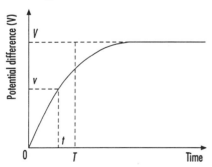

Fig. 5.22

The corresponding graph of charging current against time is shown in Fig. 5.23.

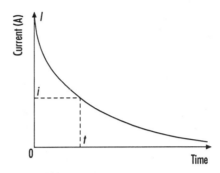

Fig. 5.23

The instantaneous values of voltage v and current i can be read respectively from the two graphs.

These values can also be found by calculation using: instantaneous p.d. (potential difference) across C is

$$v_C = V(1 - e^{-t/CR})$$

instantaneous current is

$$i = I\,e^{-t/CR}$$

and instantaneous charge is

$$q = Q(1 - e^{-t/CR})$$

5.15 TIME CONSTANT (*T*) FOR SERIES *RC* CIRCUIT

The quantity CR in the above formula is called the time constant. It determines how quickly v_R and v_C approach their final values

$$T = CR$$

Consider the point where the instantaneous time t equals the time constant T. From

$$v_C = V(1 - e^{-t/CR})$$

we have

$$v_C = V(1 - e^{-1})$$

$$= V(1 - 0.3679) = 0.6321\,V$$

From this, the time constant can be defined as the time taken for the voltage across the capacitor to rise to 0.6321 of its final steady value.

The initial growth of the capacitor voltage = $V/T = V/CR$.

Examples

1. A capacitor of 5 μF is to be charged in series with a 0.4 MΩ resistor from a 100 V supply. Calculate: (a) the time constant; (b) the initial rate of growth of capacitor voltage; (c) the initial charging current; (d) the potential difference across the capacitor and the value of the current flowing through the resistor 4 s after the supply has been switched on.

$$C = 5 \times 10^{-6} \text{ F} \quad R = 0.4 \times 10^6 \, \Omega$$

$$V = 100 \text{ V}$$

(a) Time constant is given by

$$T = CR = 5 \times 10^{-6} \times 0.4 \times 10^6 = 2 \text{ s}$$

(b) Initial rate of growth of capacitor voltage is

$$\frac{V}{CR} = \frac{100}{2} = 50 \text{ V/s}$$

(c) Initial charging current is

$$\frac{V}{R} = \frac{100}{0.4 \times 10^6} = 0.000\,25 \text{ A} = 0.25 \text{ mA}$$

(d) Potential difference across capacitor is

$$v_C = V(1 - e^{-t/CR}) = 100(1 - e^{-4/2})$$

$$= 100(1 - e^{-2})$$

$$= 100(1 - 0.1354)$$

$$= 100(0.8646) = 86.46 \text{ V}$$

Current flowing after 4 s is

$$i = I e^{-t/CR}$$

$$= 0.25 \, e^{-4/2} = 0.25 \, e^{-2}$$

$$= 0.25 \times 0.1353$$

$$= 0.033\,83 \text{ mA}$$

2. A 200 μF capacitor is connected in series with a 1000 Ω resistor across a 240 V d.c. supply. Calculate: (a) the time constant; (b) the initial rate of growth of capacitor voltage; (c) the initial charging current; (d) the time taken for the p.d. across the capacitor to grow to 120 V; (e) the current and p.d. across the capacitor 0.2 s after the supply switch is closed; (f) the energy stored in the capacitor when it is fully charged.

$$C = 200 \times 10^{-6} \text{ F} \quad R = 1000 \, \Omega$$

$$V = 240 \text{ V}$$

(a) Time constant is

$$T = CR = 200 \times 10^{-6} \times 1000$$

$$= 0.2 \text{ s}$$

(b) Initial rate of growth of capacitor voltage is

$$\frac{V}{CR} = \frac{240}{0.2} = 1200 \text{ V/s}$$

(c) Initial charging current is

$$\frac{V}{R} = \frac{240}{1000} = 0.24 \text{ A}$$

(d) From $v = V(1 - e^{-t/CR})$ we have

$$120 = 240(1 - e^{-t/0.2})$$

$$\frac{120}{240} = (1 - e^{-t/0.2})$$

$$e^{-t/0.2} = 1 - 0.5 = 0.5$$

$$e^{t/0.2} = \frac{1}{0.5} = 2.0$$

Taking logarithms to base e of both sides of the equation

$$-\frac{t}{0.2} \log_e e = \log_e 0.5$$

$$-\frac{t}{0.2} \times 1 = -0.6392$$

$$t = -0.2 \times -0.6392$$

$$t = 0.1386\,s$$

(e) From $v = V(1 - e^{-t/CR})$ V we have

$$V = 240(1 - e^{-0.2/0.2}) = 240(1 - e^{-1})$$

$$= 240(1 - 0.3679) = 240 \times 0.6321$$

$$= 151.7 \,V$$

Corresponding value of current is

$$i = I \, e^{-t/CR} \, A$$

$$= 0.24 \times e^{-1}$$

$$= 0.24 \times 0.3679 = 0.088 \, A$$

$$= 88 \, mA$$

(f) Energy stored is

$$\tfrac{1}{2}CV^2 = \tfrac{1}{2} \times 200 \times 10^{-6} \times 240^2 = 5.76 \, J$$

5.16 THE SERIES *RC* CIRCUIT – DISCHARGING

When the switch is closed in the position 'b' as shown in the circuit diagram of Fig. 5.24 the capacitor will be discharged through the resistor.

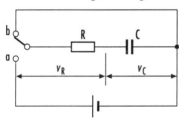

Fig. 5.24

The graph in Fig. 5.25 shows the variation of capacitor voltage with time during discharge.

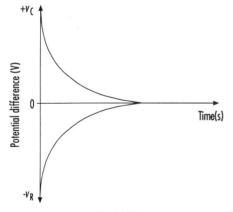

Fig. 5.25

The graph in Fig. 5.26 shows the variation of current with time during discharge.

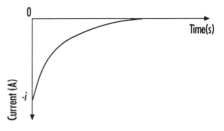

Fig. 5.26

As when charging, the time constant on discharge $T = RC$ s. However on discharge:

$$v_C = V \, e^{-t/CR}$$

$$i = -I \, e^{-t/CR}$$

$$q = Q \, e^{-t/CR}$$

$$W = \tfrac{1}{2}CV^2$$

Examples

1. An $8 \, \mu F$ capacitor is connected in series with a $50\,000 \, \Omega$ resistor across a $100 \, V$ d.c. supply. When fully charged the capacitor is disconnected from the supply and is then discharged through the resistor. Calculate: (a) the time constant of the circuit; (b) the charge of the capacitor after 0.8 s; (c) the magnitude of the current when the time is twice the time constant; (d) the magnitude of the discharge voltage when the time is twice the time constant.

$$C = 8 \times 10^{-6} \, F \quad R = 50\,000 \, \Omega$$

$$V = 100 \, V$$

(a) Time constant is

$$T = CR = 8 \times 10^{-6} \times 50\,000 = 0.4 \, s$$

(b) Total charge is

$$Q = CV = 8 \times 10^{-6} \times 100 = 0.0008 \, C$$

$$= 800 \, \mu C$$

Charge after 0.8 s is

$$q = Q \, e^{-t/CR} = 800 \, e^{-0.8/0.4} = 800 \, e^{-2}$$

$$= 800 \times 0.1353 = 108.3 \, \mu C$$

(c) Using $i = -I\,e^{-t/CR}$ current after 0.8 seconds is

$$I = \frac{V}{R} = \frac{100}{50\,000} = 0.002\,\text{A} = 2\,\text{mA}$$

Therefore

$$i = -2\,e^{-0.8/0.4}$$

$$= -0.271\,\text{mA}$$

The negative sign indicates that the current is flowing in the opposite direction to that of the charging current.

(d) Using $v = V\,e^{-t/CR}$ discharge voltage after 0.8 seconds is

$$v = 100\,e^{-0.8/0.4} = 13.54\,\text{V}$$

2. A $40\,\mu\text{F}$ capacitor is connected in series with a $100\,000\,\Omega$ resistor across a 500 V d.c. supply. Determine: (a) the time taken for the p.d. across the capacitor to rise to 100 V; (b) the time taken for the capacitor voltage to fall to 50 V after disconnecting the fully-charged capacitor from the supply and discharging it through a $0.5\,\text{M}\Omega$ resistor.

$$C = 40 \times 10^{-6}\,\text{F} \quad R = 100\,000\,\Omega$$

$$V = 500\,\text{V}$$

(a) Time constant on charge is

$$T = CR = 40 \times 10^{-6} \times 100\,000 = 4\,\text{s}$$

Using $v = V(1 - e^{-t/CR})$ we have

$$100 = 500(1 - e^{-t/4})$$

$$\frac{100}{500} = (1 - e^{-t/4})$$

$$0.2 = (1 - e^{-0.25t})$$

$$e^{-0.25t} = 1 - 0.2 = 0.8$$

Taking logarithms to base e of both sides of equation:

$$-0.25t\,\log_e e = \log_e 0.8$$

$$-0.25t \times 1 = -0.2231$$

Therefore

$$t = \frac{-0.2231}{-0.25}$$

$$= 0.8924\,\text{s}$$

(b) Time constant on discharge is

$$CR = 40 \times 10^{-6} \times 0.5 \times 10^{6} = 20\,\text{s}$$

Using $v = V\,e^{-t/CR}\,\text{V}$ we have

$$50 = 500 \times e^{-t/20}$$

$$\frac{50}{500} = e^{-0.05t}$$

$$0.1 = e^{-0.05t}$$

Taking logarithms to base e of both sides

$$-0.05t\,\log_e e = \log_e 0.1$$

$$-0.05t \times 1 = -2.3026$$

Therefore

$$t = \frac{-2.3026}{-0.05}$$

$$= 46.1\,\text{s}$$

Exercise 5.9

1. Draw a circuit that could be used for charging a capacitor using a d.c. supply.

2. Explain what is meant by the term 'time constant' in an RC circuit.

3. Sketch a graph showing the relationship between the growth of current with time during the charging period of a capacitor.

4. Sketch a graph showing the relationship between the growth of capacitor potential difference with time during the charging period of a capacitor.

5. Sketch the curves for the variation of voltage and current with time for each of the components in a series *RC* circuit when a capacitor is discharging.

6. Define the 'time constant' of a circuit that includes a resistance and capacitance connected in series.

A capacitor of $100\,\mu F$ is connected in series with an $8000\,\Omega$ resistance. Estimate the time constant of the circuit. If the combination is connected suddenly to a $100\,V$, d.c. supply, find: (a) the initial rate of rise of p.d. across the capacitor; (b) the initial charging current; (c) the ultimate charge in the capacitor.

7. A capacitor of $0.1\,\mu F$ capacitance, charged to a p.d. between plates of $100\,V$, is discharged through a resistor of $1\,M\Omega$. Calculate: (a) the initial value of the discharge current; (b) its value $0.1\,s$ later; (c) the initial rate of decay of the capacitor voltage; (d) the energy dissipated in the resistor. Using the above data, sketch the variation of discharge current with time.

8. A $100\,\mu F$ capacitor is connected in series with a $800\,\Omega$ resistor. Determine the time constant of the circuit. If the combination is connected suddenly to a $200\,V$ d.c. supply, find: (a) the initial rate of rise of p.d. across the capacitor; (b) the initial charging current; (c) the ultimate charge in the capacitor; (d) the ultimate energy stored in the capacitor.

9. A $2\,\mu F$ capacitor is joined in series with a $2\,M\Omega$ resistor to a d.c. supply of $100\,V$. Draw a current–time graph and explain what happens in the period after the circuit is made, if the capacitor is initially uncharged.

Calculate the current flowing and the energy stored in the capacitor at the end of an interval of $4\,s$ from the start.

Answers

Exercise 5.1
1. $0.001\,C$
2. $2000\,kV$, $1000\,MV/m$
3. $30\,\mu F$
4. (c) $0.01\,C$

Exercise 5.2
1. See 5.1
2. See 5.2
3. See 5.4
4. See Fig. 1.5
5. See 5.4.1 and 5.4.2
6. See 5.4.3, 5.4.4, 5.4.5, 5.4.6
7. See 5.2
8. See 5.5.1(a)
9. See 5.5.1(a)
10. See 5.5.1(b)
11. See 5.5.2
12. See 5.5.2
13. See 5.5.3 and Fig. 5.13
14. See 5.5.4
15. See 5.5.5 and 5.5.6
16. (a) Brown, Black, Brown
    (b) Brown, Red, Red
    (c) Brown, Green, Orange
    (d) Brown, Grey, Yellow
    (e) Red, Red, Green
    (f) Red, Violet, Blue
    (g) Orange, Orange, Black
    (h) Orange, White, Brown
    (i) Yellow, Violet, Red
    (j) Green, Blue, Orange
    (k) Blue, Grey, Yellow
    (l) Grey, Red, Green
17. Gold is ±5%
    Silver is ±10%
18. (a) Brown, Blue, Brown, Gold
    (b) Red, Black, Red, Gold
    (c) Orange, Blue, Yellow, Gold
    (d) Violet, Green, Green, Silver
    (e) White, Brown, Blue, Silver
    (f) Brown, Brown, Brown
    (g) Brown, Orange, Red
    (h) Red, Yellow, Orange
19. (a) $10\,000\,pF \pm 20\%$
    (b) $22\,000\,pF \pm 20\%$
    (c) $150\,000\,pF \pm 20\%$
    (d) $15\,000\,000\,pF \pm 5\%$

(e) $390\,pF \pm 5\%$
(f) $820\,000\,pF \pm 10\%$
(g) $56\,pF \pm 10\%$
(For values in μF, multiply by 10^{-6})

Exercise 5.3
1. $50\,kV/m$
2. $250\,kV/m$
3. $200\,kV/m$
4. $250\,kV/m$
5. $2\,kV/mm$
6. (b) $5\,000\,000\,V/m$

Exercise 5.4
1. (a) $0.001\,C$; (b) $125\,kV/m$; (c) $0.02\,C/m^2$
2. $0.64\,\mu F$, $8.33\,V/mm$, $0.0005\,33\,C/m^2$
3. (a) $100\,\mu C/m^2$; (b) $2400\,kV/m$

Exercise 5.5
1. (a) $11.07 \times 10^{-12}\,F$, $11.07 \times 10^{-6}\,\mu F$, $11.07\,pF$, $11.07 \times 10^{-3}\,nF$; (b) $0.002\,66\,\mu C$; (c) $0.532\,\mu C/m^2$; (d) $60 \times 10^3\,V/m$
2. $0.5\,mm$
3. $12.05\,mm$
4. $141\,m$
5. (a) Falls to $40\,V$; (b) $1.77\,\mu C/m^2$
6. (a) $20\,MV/m$; (b) $442\,\mu C/m^2$; (c) 2.5
7. $10\,\mu C/m^2$, $200\,kV/m$, $0.000\,25\,\mu F$
8. $1:2.4$, $1:2.4$
9. (a) $3\,\mu C$; (b) $0.3\,mC/m^2$; (c) 3.39
10. $4752\,V$, $0.001\,J$

Exercise 5.6
1. (a) $24.35\,nF$; (b) $9.74\,\mu C$; (c) $1000\,kV/m$; (d) $194.8\,\mu C/m^2$
2. $2212.5\,pF$

Exercise 5.7
1. $0.156\,25\,J$
2. $2\,\mu F$, $16\,\mu J$
3. (a) $177.1\,pF$; (b) $0.8854\,\mu J$; (c) $100\,kV/m$; (d) $0.8854\,\mu C/m^2$; (a) $442.7\,pF$; (b) $2.214\,\mu J$; (c) $2.214\,\mu C/m^2$
4. $5820\,V$; $0.0015\,J$

Exercise 5.8
1. (a) $63.6\,\mu\Omega$; (b) $31.8\,\mu\Omega$; (c) $127.3\,\Omega$; (d) $42.44\,\Omega$
2. (a) $12.7\,\mu F$; (b) $26.5\,\mu F$; (c) $90.9\,\mu F$; (d) $318.3\,\mu F$
3. $265.2\,\mu F$
5. $397.8\,V$
6. $9.549\,\mu F$
7. $4.819\,A$
8. $1.885\,A$
10. (a) $0.318\,H$; (b) $15.91\,\mu F$

Exercise 5.9
6. $0.8\,s$; (a) $125\,V/s$; (b) $12.5\,mA$; (c) $0.01\,C$
7. (a) $100\,\mu A$; (b) $36.8\,\mu A$; (c) $1000\,V/s$; (d) $0.5\,mJ$
8. $0.08\,s$; (a) $2500\,V/s$; (b) $0.25\,A$; (c) $0.02\,C$; (d) $2\,J$
9. $18.4\,\mu A$; $400\,\mu J$

6

COMPONENTS 3

6.1 INDUCTOR

An inductor is a device or component usually in the form of a coil, that has the property of inductance. It is this property that makes the inductor a useful component in electronics. Another name for the inductor is solenoid. The solenoid is defined as a coil of wire that has a long axial length relative to its diameter. The coil is cylindrical in form and is used to produce a known magnetic flux density along its axis. Another name for an inductor is a choke or choking coil. More detail will be provided later on these matters. The British Standard graphical symbol for an inductor can be found in Chapter 1.

6.2 FORCE ON A CURRENT-CARRYING CONDUCTOR

The magnetic field round a long straight current-carrying conductor is shown in Figs 6.1 and 6.2.

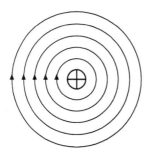

Fig. 6.1

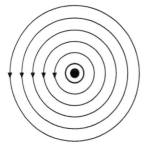

Fig. 6.2

The right hand rule identifies the direction of the lines of magnetic flux. The conductor is gripped in the right hand with the thumb pointing in the direction of current flow. The fingers then indicate the direction of the lines of magnetic flux around the conductor.

The magnetic field produced by two bar magnets having unlike poles close to one another is illustrated in Fig. 6.3.

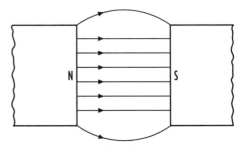

Fig. 6.3

The interaction of two magnetic fields is illustrated in Figs 6.4 and 6.5.

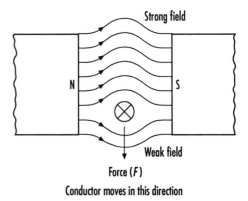

Strong field

Weak field

Force (*F*)

Conductor moves in this direction

Fig. 6.4

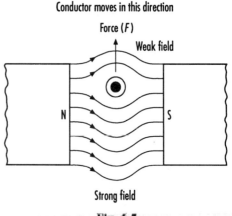

Conductor moves in this direction

Force (*F*)

Weak field

Strong field

Fig. 6.5

The force (F) on the conductor is directly proportional to the current (I) and the length (l) of the conductor in the magnetic field.

$$F \propto lI$$

$$F = \text{constant} \times lI$$

The constant factor is the magnetic flux density (B), measured in tesla (T), of the uniform magnetic field. Therefore

$$F = BlI$$

In electrical engineering the effect has many applications, e.g. an electric motor, a device which converts electrical energy to mechanical energy. One magnetic field is produced in a motor by a fixed electromagnet called the stator, and the other by the electromagnet called the rotor which is free to move. Other applications for this effect can be observed in moving coil instruments, cathode ray tubes and loud-speaker coil systems.

Examples

1. A copper wire carrying a current of 50 mA lies within, and at right angles to a uniform magnetic field 40 mm wide in the direction of the wire. Calculate the force produced when the flux density is 1.2 T.

    $$B = 1.2\,\text{T} \quad l = 40\,\text{mm} = 40 \times 10^{-3}\,\text{m}$$

    $$I = 50\,\text{mA} = 50 \times 10^{-3}\,\text{A}$$

    Using $F = BlI$

    $$F = 1.2 \times 40 \times 10^{-3} \times 50 \times 10^{-3}$$

    $$= 0.0024\,\text{N}$$

2. A conductor 0.7 m long is carrying a current of 50 A at right angles to a magnetic field. If the force on the wire is 4 N determine the density of the magnetic field.

    $$l = 0.7\,\text{m} \quad I = 50\,\text{A} \quad F = 4\,\text{N}$$

    Using $F = BIl$

    $$B = \frac{F}{lI} = \frac{4}{0.7 \times 50} = 0.1143\,\text{T}$$

3. A current of 3 A is passing through 400 conductors in an electric motor armature. Each conductor is 150 mm long and lying in and at right angles to a magnetic field of flux density 0.8 T. If each conductor is situated 25 mm from the axis of rotation calculate: (a) the force produced; (b) the torque produced.

    $$B = 0.8\,\text{T} \quad l = 150 \times 10^{-3}\,\text{m}$$

    $$I = 3\,\text{A} \quad r = 25 \times 10^{-3}\,\text{m}$$

    (a) From $F = BlI$

    $$F = 0.8 \times 150 \times 10^{-3} \times 400 \times 3$$

    $$= 144\,\text{N}$$

    (b) Torque $= F \times$ radius

    $$= 144 \times 25 \times 10^{-3}$$

    $$= 3.6\,\text{Nm}$$

Exercise 6.1

1. Draw a diagram showing a current-carrying conductor lying in, and at right angles to, a uniform magnetic field. State the expression for the force acting on the conductor.

    A conductor 0.1 m long lies in, and at right angles to a uniform magnetic field of density 0.5 T. Calculate the force on the conductor when the current is 100 A.

2. Explain what happens when a long straight conductor is moved through a uniform magnetic field at constant velocity. Assume that the conductor moves perpendicular to the field. If the ends of the conductor are connected together through an ammeter, what will happen?

    A conductor 0.7 m long is carrying a current of 75 A and is placed at right angles to a magnetic field of uniform flux density. Calculate the flux density of the magnetic field if the mechanical force acting on the conductor is 26 N.

3. Write down an expression for the force acting on a conductor of length l carrying a current I, lying in and at right angles to a magnetic field of flux density B.

    A coil has a mean width of 40 mm and a mean axial length of 50 mm. There are 100 turns of wire and the current flow is 0.03 A. If the coil is lying at right angles to a magnetic field of uniform flux density 0.45 T, find the force acting on the coil, and the maximum torque on the coil in Nm.

4. A current-carrying conductor is placed at right angles to a magnetic field. Show clearly the direction in which the force exerted on the conductor acts for a given direction of the field and the current.

    If the conductor is 0.9 m long and the field is of uniform density 0.8 T, find the force exerted on the conductor when the current is 15 A.

    Show briefly how this principle is used in the moving-coil instrument.

5. Draw a diagram to show clearly the relative directions of mechanical force, magnetic field, and current for a current-carrying conductor situated at right angles to a magnetic field. How could the direction of force be reversed?

    Each conductor on the armature of a d.c. motor carries a current of 30 A and has an effective length of 0.3 m. Calculate the magnitude of the force exerted on such a conductor when it lies in a magnetic field of density 0.95 T.

    If the conductor acts at a radius of 0.2 m determine the torque which it exerts on the armature.

6. State the expression for the force acting on a current-carrying conductor lying in, and at right angles to, a magnetic field.

    A rectangular former has a mean width of 15 mm and a mean axial length of 20 mm. It is wound with 50 turns of wire and is situated in an air gap of uniform flux density 0.2 T.

    When the current in the coil is 0.01 A, calculate: (a) the force acting on one side of the coil; (b) the total torque on the coil.

7. A straight horizontal wire carries a steady current of 150 A and is situated in a uniform magnetic field of 0.6 T acting vertically downwards. Determine the magnitude of the force in newtons per metre length of conductor and the direction in which it acts.

    Explain how the principle of the force on a current-carrying conductor situated in a magnetic field is utilized in order to obtain rotation of the armature of a d.c. motor.

6.3 LAWS OF ELECTROMAGNETIC INDUCTION

Faraday discovered that when movement takes place between a conductor and a magnetic field, an electromotive force is induced in the conductor.

Neumann showed that when a conductor cuts a magnetic field the induced e.m.f. is proportional to the rate at which the flux changes.

Lenz showed that the induced e.m.f. is in a direction such as to oppose the change producing it.

6.4 MAGNITUDE OF INDUCED E.M.F.

Consider a current-carrying conductor in a magnetic field as illustrated in Fig. 6.4. The force to move the conductor downward is

$$F = BlI$$

and

$$\text{work done} = \text{force} \times \text{distance moved}$$

$$= BlI \times d \text{ newton} - \text{metre}$$

Electrical power generated $= EI$ and electrical energy $= EIt$ joules. It is known that 1 Nm = 1 J. Therefore

$$EIt = BlId$$

and

$$E = \frac{BlId}{It} = \frac{Bld}{t}$$

But velocity $(v) = \text{distance} (d)/\text{time} (t)$. Therefore

$$E = Blv$$

The expression $E = (Bld)/t$ can also be modified as follows. Let a be the area of flux covered by the conductor, ld. Then

$$E = \frac{Ba}{t}$$

and the magnetic flux Φ is given by

$$\Phi = Ba$$

so

$$E = \frac{\Phi}{t}$$

The formula

$$E = \frac{\Phi}{t}$$

applies in the case of a single conductor; for a coil with N turns

$$E = \frac{\Phi N}{t}$$

If the conductor is moving at right angles to the magnetic field we have $E = Blv$ but when the conductor is moving at an angle to the magnetic field we have $E = Blv \sin \theta$ where $\theta = $ angle between the direction of the magnetic field and the direction of movement as illustrated in Fig. 6.6.

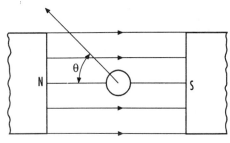

Fig. 6.6

The expressions

$$E = Blv$$

and

$$E = \frac{\Phi N}{t}$$

do not always take into account that according to Lenz's law the induced e.m.f. acts in opposition to the change which produces it. The expressions should be written

$$E = -Blv$$

and

$$E = \frac{-\Phi N}{t}$$

where the negative sign implies the direction of the induced e.m.f. However, when solving problems the signs can often be ignored.

Examples

1. A straight conductor 0.5 m long moves at right angles to a magnetic field of flux density 0.8 T at a velocity of 10 m/s. Calculate: (a) the e.m.f. induced in the conductor; (b) the force acting on the conductor if the conductor has a resistance of 0.2 Ω.

$$B = 0.8\,\text{T} \quad l = 0.5\,\text{m}$$

$$v = 10\,\text{m/s} \quad R = 0.2\,\Omega$$

(a) Using $E = Blv$

$$E = 0.8 \times 0.5 \times 10$$

$$= 4\,\text{V}$$

(b) Using $I = E/R$

$$I = \frac{4}{0.2} = 20\,\text{A}$$

From $F = BlI$

$$F = 0.8 \times 0.5 \times 20$$

$$= 8\,\text{N}$$

2. Calculate the induced e.m.f. in a coil of 5000 turns which is linked with a magnetic flux which changes from 0.04 to 0.01 weber (Wb) in 2 s.

$$N = 5000 \quad t = 2\,s$$

$$\Phi = 0.04 - 0.01 = 0.03\,Wb$$

Using $E = (\Phi N)/t$

$$E = \frac{0.03 \times 5000}{2}$$

$$= 75\,V$$

3. Determine the rate of change of flux that is required to induce an e.m.f. of 20 000 V in a motor car ignition coil consisting of 10 000 turns.

$$E = 20\,000\,V \quad N = 10\,000 \quad t = 1\,s$$

Using $E = \Phi N /t$

$$\Phi = \frac{Et}{N} = \frac{20\,000 \times 1}{10\,000} = 2\,Wb/s$$

Exercise 6.2

1. Draw diagrams showing the magnetic field in the vicinity of: (a) a single conductor; (b) a solenoid. Show clearly in each diagram the direction of the magnetic field and current. Indicate the correct polarity of the solenoid.

    A straight conductor 0.5 m long (which forms part of a closed circuit) carries a current of 20 A and lies at right angles to a uniform magnetic field of 0.25 T.

    Calculate: (a) the force, in newtons, acting on the conductor; (b) the e.m.f. induced in the conductor when it is moved against the force at a uniform velocity of 12 m/s; (c) the power in watts required to move the conductor.

    Show that the electrical power produced is equal to the mechanical power producing motion.

2. (a) State (i) Faraday's Law of electromagnetic induction, and (ii) Lenz's Law.
    (b) Calculate the e.m.f. generated in the axles of a railway train when travelling at 100 km/h. The axles are 1.4 m in length and the component of the earth's magnetic field is $40\,\mu T$.

3. Specify the laws of electromagnetic induction and give one application of each law.

    A conductor 0.1 m in length moves with uniform velocity of 2 m/s at right angles both to itself and to a uniform magnetic field having a flux density of 1 T. Calculate the induced e.m.f. between the ends of the conductor.

    Draw a diagram giving a cross-section of the conductor and mark on it the direction of the induced e.m.f., the direction of motion and the direction of the field. Describe the method whereby the direction of the induced e.m.f. is determined.

4. State Faraday's Law of electromagnetic induction.

    A straight conductor moves through a magnetic field at right angles to the field. Show by means of a diagram the direction of the induced e.m.f. relative to the field and the motion of the conductor.

    A straight axial conductor 0.25 m long is fixed on the surface of a cylindrical armature of diameter 0.4 m. If the armature rotates at 1000 rev/min and the mean gap flux density at the conductor is 0.6 T, find the e.m.f. induced in the conductor.

5. State Lenz's Law.

    A conductor 0.5 m long is moved at a uniform speed at right angles to its length and to a uniform magnetic field having a density of 0.4 T.

    If the e.m.f. generated in the conductor is 2 V and the conductor forms part of a closed circuit having a resistance of 0.5 Ω, calculate: (a) the velocity of the conductor in m/s; (b) the force acting on the conductor in newtons; (c) the work done in joules when the conductor has moved 0.6 m.

6. (a) Define the ampere in terms of the force between two parallel conductors. (b) A straight conductor 0.2 m long is moved at a uniform speed of 6 m/s at right angles to its length and to a uniform magnetic field. Calculate the flux density of the field if the e.m.f. induced in the conductor is 1.8 V. If this conductor forms part of a closed circuit having a total resistance of 0.1 Ω, calculate the force on the conductor in newtons.

7. A conductor 0.2 m long is moved at a uniform velocity of 40 m/s at right angles to its length and to a magnetic field. Find the e.m.f. induced between the ends of the conductor if the density of the magnetic field is 1.2 T.

If the ends of this conductor are joined through an external resistance so that a current of 50 A flows, find the retarding force. Find also the power required to keep the conductor moving.

8. Show that the e.m.f. generated in a conductor l metres long rotating in a magnetic field of flux density B tesla at a velocity of v metres/second is given by the expression $E = Blv \sin \theta$ volts.

A square coil 0.1 m × 0.1 m consists of 20 turns and is rotated about an axis through the centre of uniform flux density 0.5 T. Calculate the maximum value of the e.m.f. induced in the coil when it is rotated at a speed of 3000 rev/min.

9. A length of stiff copper wire is made into a rectangle measuring 0.24 m by 0.1 m. It is rotated at a uniform speed of 600 rev/min about one of its longer sides as an axis, this side lying in and at right angles to a magnetic field of 0.5 T. Calculate the average e.m.f. induced in the coil.

6.5 SELF INDUCTANCE OF A COIL

A coil has a self inductance (L) of 1 henry (H) if an e.m.f. of 1 V is induced in the coil when the current through the coil changes at the rate of 1 A/s.

$$L = \frac{E}{(I/t)} = \frac{Et}{I}$$

But

$$E = \frac{\Phi N}{t}$$

so

$$L = \frac{\Phi N}{t} \times \frac{t}{I} = \frac{\Phi N}{I}$$

Examples

1. A magnetic flux of 400 μWb is produced by a current of 5 A passing through a coil of 600 turns. Calculate: (a) the inductance of the coil; (b) the e.m.f. induced in the coil when the current is reversed in 0.3 s.

$$\Phi = 400 \ \mu\text{Wb} = 400 \times 10^{-6} \ \text{Wb}$$

$$N = 600 \quad I = 5 \, \text{A}$$

(a) Using $L = \Phi N / I$

$$L = \frac{400 \times 10^{-6} \times 600}{5} = 0.048 \, \text{H}$$

The flux is changing from +400 to −400 μWb in 0.3 s so $\Phi = 800 \, \mu$Wb.

(b) Using $E = \Phi N / t$

$$E = \frac{800 \times 10^{-6} \times 600}{0.3} = 1.6 \, \text{V}$$

2. If a coil whose self inductance is 0.6 H carries a current which changes at a rate of 10 mA every 2 s, what value of e.m.f. will be induced?

$$L = 0.6 \, \text{H} \quad I = 10 \, \text{mA} = 10 \times 10^{-3} \text{A}$$

$$t = 2 \, \text{s}$$

Using $L = Et / I$

$$E = \frac{LI}{t}$$

$$= \frac{0.6 \times 10 \times 10^{-3}}{2}$$

$$= 0.003 \, \text{V} = 3 \, \text{mV}$$

Exercise 6.3

1. Explain what is meant by the term self inductance. Give the name and define the unit of self inductance.

   A coil of 500 turns produces a flux of 1.5 mWb when carrying a current of 3 A. What is its inductance? What e.m.f. would be induced in the coil if the current were reduced uniformly to zero in 0.01 s?

2. Explain the meaning of self inductance and define the unit in which it is measured.

   A coil consists of 750 turns, and a current of 10 A in the coil gives rise to a magnetic flux of 1200 μWb. Calculate the inductance of the coil, and determine the average e.m.f. induced in the coil when this current is reversed in 0.01 s. Neglect the resistance of the coil. Determine also the energy stored.

3. If a coil of 150 turns is linked with a flux of 0.01 Wb when carrying a current of 10 A, calculate the inductance of the coil. If this current is uniformly reversed in 0.1 s, calculate the induced e.m.f.

4. (a) A coil of 1000 turns, when carrying a current of 2 A, sets up a magnetic flux of 0.05 Wb. If the current is reversed in $\frac{1}{100}$ s, what is the average e.m.f. induced in the coil? (b) An iron-cored coil of 2000 turns produces a flux of 0.008 Wb. When the switch in the circuit is opened the flux falls to 0.001 Wb in 0.02 s. What is the average e.m.f. induced?

5. Explain, with the aid of a sketch, what is meant by the term self inductance. Define the term in which it is measured.

   If a coil of 200 turns is linked with a flux of 0.01 Wb when carrying a current of 10 A, calculate the inductance of the coil. If this current is reduced to zero in 0.05 s, calculate the induced e.m.f.

6. What is understood by the term self inductance?

   An air cored coil of 2000 turns produces a flux of 1 mWb when a current of 1.5 A is flowing. If the current is reduced uniformly to 0.5 A in 0.1 s, calculate the induced voltage in the coil.

7. (a) Draw a diagram to show the magnetic field set up by a current flowing in a straight conductor and hence explain the production of a force between two parallel current-carrying conductors. (b) Explain the features of a coil which would possess a high value of inductance. (c) If a coil whose self inductance is 0.5 H carries a current which changes at a rate of 10 mA/s, what value of e.m.f. will be induced?

8. (a) Define the unit of inductance in terms of (i) flux linkages, (ii) current change. (b) The inductance between two windings of an induction coil is 10 H. A current of 2 A in the primary is reduced to zero in 0.001 s. Calculate the average value of the induced e.m.f. in the secondary winding during this time.

9. Name and define the unit of self inductance.

   A large electromagnet is wound with 1000 turns. A current of 2 A in this winding produces a flux through the coil of 0.03 Wb. Calculate the inductance of the electromagnet.

   If the current in the coil is reduced from 2 A to zero in 0.1 s, what average e.m.f. will be induced in the coil?

10. (a) A coil of 1500 turns is linked by a magnetic flux of 600 Wb. Calculate the e.m.f. induced in the coil when (i) the flux is reversed in 0.001 s, and (ii) the flux is reduced to zero in 0.1 s. (b) State a practical example of an e.m.f. being induced when a circuit is broken, which could be very dangerous. (c) What precautions would be taken to minimize the danger of such an e.m.f.?

11. An iron-cored coil of 2000 turns produces a magnetic flux of 30 mWb when a current of 10 A is flowing from the d.c. supply. Find the average value of induced e.m.f. if the time of opening the supply switch is 0.12 s. The residual flux of the iron is 2 mWb.

    What would be the inductance of this circuit under these conditions?

6.6 ENERGY

Energy can neither be created nor destroyed, therefore when the switch in an inductive circuit is opened, the current flow will cease and the magnetic energy stored will be dissipated as heat, either through a discharge resistor or in an arc at the switch contacts. Energy is given by

$$W = \text{power} \times \text{time} = E \times \text{current} \times t$$

$$= \frac{LI}{t} \times \text{current} \times t = LI \times \text{current}$$

If we assume that the current changes from a maximum value to zero in time (t) seconds then the mean current is $I/2$. Therefore

$$W = LI \times \frac{I}{2} = \tfrac{1}{2}LI^2$$

Example

1. Calculate the energy stored in a magnetic field of a coil which has a resistance of 20 Ω and an inductance of 2 H. The supply e.m.f. is 50 V.

    $$R = 20\,\Omega \quad L = 2\,H \quad E = 50\,V$$

    Using $I = E/R$

    $$I = \frac{50}{20} = 2.5\,A$$

    Using $W = \tfrac{1}{2}LI^2$

    $$W = \tfrac{1}{2} \times 2 \times 2.5^2 = 6.25\,J$$

Exercise 6.4

1. A coil of resistance 5 Ω has an inductance of 2 H. Calculate: (a) the energy stored in the coil when working off a 200 V d.c. supply; (b) the current flowing through the coil.

2. A coil has an inductance of 0.5 H and takes a current of 2 A from a 12 V supply. Calculate: (a) the energy stored in the coil; (b) the resistance of the coil.

3. A coil of wire of self inductance 1.5 H has a store of energy of 48 J from a supply of 12 V. Calculate: (a) the resistance of the coil; (b) the current flowing through the coil.

4. The coil of an electromagnet has a cross-sectional area of 2 mm² and is wound with 300 turns. Calculate the energy stored in the magnetic field when the current in the coil is 5 A and produces a flux density of 1.2 T.

5. A coil of resistance 200 Ω is connected to a d.c. supply. If the power taken is 225 W and the energy stored is 0.04 J, calculate the supply voltage and the inductance of the coil.

6. A coil is wound with 400 turns and the cross-sectional area of the magnetic path is 5 mm². Calculate the energy stored in the coil when a current of 5 A produces a flux density of 1.4 T.

7. (a) Define the unit of self inductance. (b) An electromagnet is wound with 200 turns. A current is 0.4 A in this winding produces a flux through the coil of 0.03 Wb. Calculate: (i) the inductance of the electromagnet, (ii) the energy stored in the inductor, (iii) the average value of induced e.m.f. in the coil, if the current is reduced to zero in 0.1 s.

8. An air-cored coil of 1000 turns carrying a current of 4 A produces a flux of 0.01 Wb linking the coil. Calculate the coefficient of self inductance of the coil, and determine also the stored energy.

9. A coil is wound on a non-magnetic ring of uniform cross-sectional area 400 mm² and having a mean circumference of 0.25 m. Calculate: (i) the number of turns necessary to produce an inductance of 10 mH, (ii) the energy stored when a current of 2 A flows through the inductor.

6.7 TYPES OF INDUCTOR

6.7.1 Fixed inductor

This type of inductor has the same shape as that of a carbon-composition resistor. A typical size is 10 mm long, 4 mm diameter. They are available from 0.1 μH to about 4700 μH. They are constructed by being moulded and have high reliability and give a good electrical performance with a tolerance of $\pm$10%. They operate in the temperature range -55 to $+125°C$ and have a dielectric strength in the area of 700 V. The small size of this component makes it suitable as a radio frequency inductor and for low power switching regulator applications.

6.7.2 Moulded chip inductor

The outline shape for this inductor is shown in Fig. 6.7. They are available from 1 μH to 1000 μH with a 10% or 20% tolerance. Because of the moulding techniques used this makes them very strong and moisture-resistant. They operate in the temperature range -40 to $+105°C$.

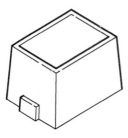

Fig. 6.7

6.7.3 High-current bobbin core inductor

The shape shown in Fig. 6.8 is cylindrical with a typical length of 16 mm and diameter 24 mm. They are available from 22 μH to 2200 μH with a tolerance of $\pm$10%. The core is manufactured from a high-saturation flux density material with the winding being insulated by a heat-shrunk sleeve. The inductor may be chassis-mounted or alternatively mounted on a PCB. They are often used in switching regulators, filter and power line applications.

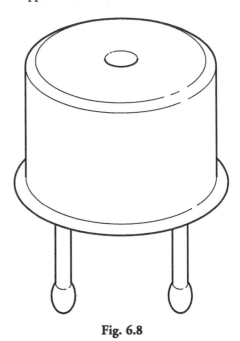

Fig. 6.8

6.7.4 HIGH-FREQUENCY SURFACE-MOUNTED INDUCTOR

This type of inductor is constructed by using thin film multi-layer technology. The terminations are nickel-solder coated and are compatible with automatic soldering techniques.

They are rectangular in shape with typical dimensions of 2 mm × 1.5 mm × 1 mm. The inductance range available is from 2.5 nH to 25 nH. The main features of this inductor are a low d.c. resistance, rugged construction, very low profile and a tight tolerance. It has uses in mobile communications, satellite TV receivers, vehicle location systems, filter and matching networks.

6.8 PURE INDUCTANCE ON A.C.

The circuit diagram is shown in Fig. 6.9.

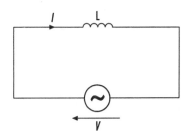

Fig. 6.9

In an inductive circuit without any resistance the current lags behind the voltage by 90°, or

$$\left(\frac{\pi}{2} \text{ radians} \right)$$

as shown in Fig. 6.10.

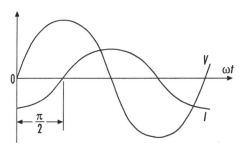

Fig. 6.10

The phasor diagram for this situation is shown in Fig. 6.11.

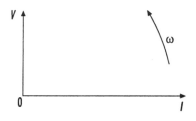

Fig. 6.11

The opposition provided by an inductor to the alternating current is called inductive reactance (X_L)

$$X_L = \frac{V}{I}$$

The inductive reactance can be found by using the inductance (L) and the frequency (f) on the a.c. supply, i.e.

$$X_L = 2\pi f L = \omega L$$

The graphs shown in Figs 6.12 and 6.13 show the relationship between the current and frequency and between the current and frequency and between the inductive reactance and frequency for an inductive circuit.

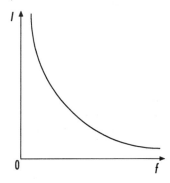

Fig. 6.12

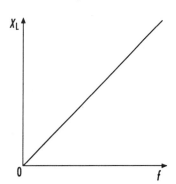

Fig. 6.13

Examples

1. Calculate the inductive reactance of a coil of 0.1 H inductance when connected to a 50 Hz supply.

$$L = 0.1\,\text{H} \quad f = 50\,\text{Hz}$$

Using $X_L = 2\pi f L$

$$X_L = 2 \times 3.142 \times 50 \times 0.1 = 31.42\,\Omega$$

2. Calculate the inductance of a coil taking a current of 4 A when connected to a 240 V 50 Hz supply.

$$V = 240\,\text{V} \quad I = 4\,\text{A} \quad f = 50\,\text{Hz}$$

Using $X_L = V/I$

$$X_L = \frac{240}{4} = 60\,\Omega$$

From $X_L = 2\pi f L$

$$L = \frac{X_L}{2\pi f}$$

$$= \frac{60}{2 \times 3.142 \times 50} = 0.191\,\text{H}$$

Exercise 6.5

1. Calculate the inductive reactance of a coil having an inductance of 0.01 H when connected to a 50 Hz supply.

2. Determine the inductance of a coil connected to a 100 V 50 Hz supply when a current of 2 A is flowing.

3. A coil having an inductance of 0.2 H and negligible resistance is connected across a 200 V a.c. supply. Calculate the current if the frequency is: (a) 25 Hz, (b) 500 Hz.

4. A coil of pure inductance causes a voltage drop of 96 V when the current flowing through it is 2.4 A at 50 Hz. Calculate the inductive reactance and coil inductance.

5. Complete the following table:

Inductance	0.4 H			0.2 H
Applied voltage		120 V	12 V	
Current	2 A	10 A		0.5 A
Frequency	50 Hz	25 Hz	100 Hz	
Inductive reactance			6 Ω	20 Ω

6. An inductive reactor gives a current flow of 12 A from a 240 V 50 Hz supply. Calculate the current which will flow at the constant supply voltage if the frequency changes to: (a) 35 Hz; (b) 65 Hz.

7. The inductance of a coil at a constant frequency of 50 Hz, can be varied from 0.1 to 0.05 H. Plot a graph to show how inductive reactance varies with change in inductance.

8. A coil of 0.05 H inductance is connected to a constant supply of 100 V. The frequency can be varied from 20 Hz to 60 Hz. Plot a graph to show how the current flowing through the inductor changes according to the frequency.

9. Define the term 'inductive reactance'.

10. When an inductor of 10 Ω inductive reactance and negligible resistance is connected across a 50 Hz supply, the current flowing is 4 A. Draw a graph to scale to show the variation of inductive reactance with frequency from 25 Hz to 100 Hz in steps of 25 Hz. Also draw a graph to scale to show the variation of current for the same frequency range on constant supply voltage. (Make the horizontal axis the frequency scale.)

6.9 PURE INDUCTANCE ON D.C.

When an inductor is connected to a direct supply, there will be a current growth, which theoretically goes on forever, but after a time it is difficult to note any change in growth.

The opposite to current growth is current decay. This can take place when the supply when the supply to the inductor is shorted out. The current will then drop away, i.e. decay, to virtually nothing. Graphs of growth and decay can be drawn as shown in the following sections.

6.10 THE SERIES *RL* CIRCUIT – CURRENT GROWTH

When the switch is closed as shown in the circuit diagram of Fig. 6.14 the current does not reach its final value immediately but rises with time as shown in Fig. 6.15.

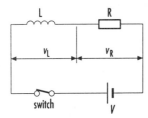

Fig. 6.14

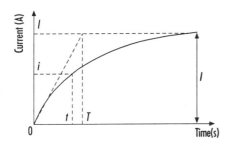

Fig. 6.15   Growth of current in an inductive circuit.

The shape of the graph obtained for the variation of voltage with time during the growth of current is shown in Fig. 6.16.

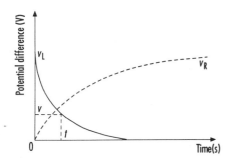

Fig. 6.16   Voltage variation with time.

From each of these graphs, the instantaneous values of voltage v and current i, for a given time, can be taken. These values can also be found by calculation using instantaneous current

$$i = I(1 - e^{-Rt/L})$$

instantaneous voltage

$$v_L = V e^{-Rt/L}$$

and instantaneous p.d. across R

$$v_R = V(1 - e^{-Rt/L})$$

6.11 TIME CONSTANT (*T*) FOR SERIES *RL* CIRCUIT

The quantity L/R is called the time constant and the initial rate of growth of current is V/L. Consider the point where the instantaneous time t equals the time constant T.
From $i = I(1 - e^{-Rt/L})$ we have

$$i = I(1 - e^{-t/(L/R)}) = I(1 - e^{-t/t})$$

so

$$i = I(1 - e^{-1})$$
$$= I(1 - 0.3679) = 0.6321I$$

From this the time constant can be defined as the time taken for the current to grow to 0.6321 of its final value.

Examples

1. A coil has a self inductance of 2.5 H and a resistance of 50 Ω. It is connected across a 200 V d.c. supply. Calculate: (a) the time constant for the circuit; (b) the rate of growth of current at the instant of closing the switch, (c) the final value of the steady current; (d) the value of the current 0.1 s after the switch is closed; (e) the induced e.m.f. in the inductor 0.15 s after the switch is closed; (f) the potential difference across the resistor 0.15 s after the switch is closed; (g) the amount of energy stored.

$L = 2.5\,\text{H} \quad R = 50\,\Omega \quad V = 200\,\text{V}$

(a) Time constant

$$T = \frac{L}{R} = \frac{2.5}{50} = 0.05\,\text{s}$$

(b) Initial rate of current growth

$$\frac{V}{L} = \frac{200}{2.5} = 80\,\text{A/s}$$

(c) Final value of steady current

$$I = \frac{V}{R} = \frac{200}{50} = 4\,\text{A}$$

(d) Value of current

$$i = I(1 - e^{-Rt/L})$$
$$= 4(1 - e^{-0.1/0.05})$$
$$= 4(1 - e^{-2}) = 3.46\,\text{A}$$

(e) Induced e.m.f.

$$v_L = V e^{-Rt/L}$$
$$= 200\,e^{-0.15/0.05} = 200\,e^{-3}$$
$$= 200 \times 0.0498$$
$$= 9.96\,\text{V}$$

(f) Potential difference across the resistor

$$v_R = V(1 - e^{-Rt/L})$$
$$= 200(1 - e^{-0.15/0.05})$$
$$= 200(1 - e^{-3}) = 200(1 - 0.0498)$$
$$= 200 \times 0.9502 = 190\,\text{V}$$

(g) Energy stored

$$W = \tfrac{1}{2}LI^2 = \tfrac{1}{2} \times 2.5 \times 4^2 = 20\,\text{J}$$

2. A coil having a resistance of $20\,\Omega$ and an inductance of $0.4\,\text{H}$ is connected across a $240\,\text{V}$ d.c. supply. Determine: (a) the rate of change of current at the instant of closing the switch; (b) the final steady value of the current; (c) the time constant of the circuit; (d) the time taken for the current to rise to 50% of its final value; (e) the value of the stored energy.

$R = 20\,\Omega \quad L = 0.4\,\text{H} \quad V = 240\,\text{V}$

(a) Initial rate of change of current

$$\frac{V}{L} = \frac{240}{0.4} = 600\,\text{A/s}$$

(b) Final steady value of current

$$I = \frac{V}{R} = \frac{240}{20} = 12\,\text{A}$$

(c) Time constant

$$T = \frac{L}{R} = \frac{0.4}{20} = 0.02\,\text{s}$$

(d) Using $i = I(1 - e^{-Rt/L})$

$$6 = 12(1 - e^{-20t/0.4})$$
$$\frac{6}{12} = (1 - e^{-50t})$$
$$e^{-50t} = 1 - 0.5 = 0.5$$

Taking logarithms to base e of both sides of the equation:

$$-50t \log_e e = \log_e 0.5$$
$$-50t \times 1 = -0.6931$$
$$t = \frac{-0.6931}{-50}$$
$$t = +0.01386\,\text{s}$$

(e) Stored energy

$$W = \tfrac{1}{2}LI^2$$
$$= \tfrac{1}{2} \times 0.4 \times 12^2 = 28.8\,\text{J}$$

6.12 THE SERIES *RC* CIRCUIT – CURRENT DECAY

For current decay to take place a switch is placed across the cell of the circuit used in Fig. 6.14. With this additional switch closed the external voltage is zero. The decay curves of voltage and current against time are shown in Figs 6.17 and 6.18.

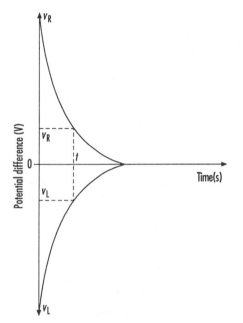

Fig. 6.17

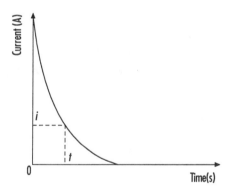

Fig. 6.18

As for current growth the time constant $T = L/R$ but on decay

$$i = I e^{-Rt/L}$$

$$v_R = V e^{-Rt/L}$$

$$v_L = -V e^{-Rt/L}$$

Examples

1. A coil has a resistance of $250\,\Omega$ and an inductance of $5\,\text{H}$. The coil is connected across a $100\,\text{V}$ d.c. supply and allowed to reach its final steady current. At this point the supply is removed and the coil is simultaneously short-circuited. Calculate the value of decay current $0.04\,\text{s}$ after the supply is disconnected.

$$R = 250\,\Omega \quad L = 5\,\text{H} \quad V = 100\,\text{V}$$

Using $i = I e^{-Rt/L}$

$$i = \frac{100}{250}\, e^{-250 \times 0.04/5} = 0.4\, e^{-2}$$

$$= 0.4 \times 0.1353 = 0.054\,12\,\text{A}$$

2. A coil has an inductance of $0.225\,\text{H}$ and a resistance of $60\,\Omega$ and is connected across a d.c. supply of $100\,\text{V}$. Determine: (a) the e.m.f. induced in the coil $0.003\,\text{s}$ after the switch is closed; (b) the time for the induced e.m.f. to change from $-100\,\text{V}$ to $-20\,\text{V}$.

$$L = 0.225\,\text{H} \quad R = 60\,\Omega \quad V = 100\,\text{V}$$

(a) The instantaneous e.m.f. induced

$$v = -V e^{-Rt/L}$$

$$= -100 \times e^{-(60 \times 0.003)/0.225}$$

$$= -100 e^{-0.8}$$

$$= -100 \times 0.449$$

$$= -44.9\,\text{V}$$

(b) Using $v = -V e^{-Rt/L}$

$$-20 = -100 \times e^{-60t/0.225}$$

$$\frac{-20}{-100} = e^{-267t}$$

$$0.2 = e^{-267t}$$

Taking logarithms to base e of both sides:

$$\log_e 0.2 = -267t \log_e e$$

$$-1.609 = -267t \times 1$$

$$t = \frac{-1.609}{-267}$$

$$t = 0.006035$$

Exercise 6.6

1. A coil with a self inductance of 2.4 H and resistance 12 Ω is suddenly switched across a 120 V d.c. supply of negligible internal resistance. Determine the time constant of the coil, the instantaneous value of the current after 0.1 s, the final steady value of the current, and the time taken for the current to reach 5 A.

2. A coil having a resistance of 250 Ω and an inductance of 20 H is connected across a 50 V d.c. supply. Calculate: (a) the initial rate of growth of the current; (b) the value of the current after 0.1 s; (c) the time required for the current to grow to 0.1 A.

3. The field winding of a d.c. machine has an inductance of 10 H and takes a final current of 2 A when connected to a 200 V d.c. supply. Calculate: (a) the initial rate of growth of current; (b) the time constant; (c) the current when the rate of growth is 5 A/s.

4. A relay has a resistance of 300 Ω and is switched on to a 110 V d.c. supply. If the current reaches 63.2% of its final steady value in 0.002 s determine: (a) the time constant of the circuit; (b) the inductance of the circuit; (c) the final steady value of the current; (d) the initial rate of rise of current.

5. A constant voltage V is maintained across an inductance of L hertz in series with a resistance of R ohms. Write down an expression for the current that flows in the circuit t s after switching on.

   On what factors does the rate of rise of current depend?

   What is the initial rate of rise of current?

   A relay coil of resistance 200 Ω and inductance 8 H is connected in series with a 100 Ω resistor and 60 V battery. The relay operates when the current in its coil is 31.6 mA. How long does it take to operate?

   The operation of the relay armature increases the inductance of the coil to 20 H. Sketch the current/time curve from the moment of switch-on, showing the effect of this increase in inductance.

   How much energy is stored in the magnetic field when the current has reached a constant value?

6. A relay has a coil resistance of 20 Ω and an inductance of 0.5 H. It is energized by a direct voltage pulse which rises from 0–10 V instantaneously, remains constant for 0.25 s, and then falls instantaneously to zero. If the relay contacts close when the current is 200 mA (increasing) and open when it is 100 mA (decreasing) find the total time during which the contacts are closed.

7. A coil has a self inductance of 2 H and a d.c. resistance of 200 Ω. It is switched suddenly across a 100 V d.c. supply of negligible internal resistance. Sketch the curve of current plotted against time and calculate: (a) the rate of rise of current in amperes per second at the instant of switching on; (b) the value of the final steady current.

8. A coil of 2000 turns has a resistance of 50 Ω and produces a magnetic flux of 5×10^{-3} Wb when a steady current of 4 A is passed through it. The coil is connected across a d.c. supply of 200 V. Calculate: (a) the time constant of the coil; (b) the induced e.m.f. in the coil at the instant when the current has risen to 3 A; (c) the rate of growth of the current at this same instant; (d) the energy stored in the magnetic field when the current has reached its steady value.

6.13 TRANSFORMERS

6.13.1 Transformer construction and action

A transformer is a device without moving parts; it consists of a steel former, a primary winding, and a secondary winding.

Common types of former are shown in Figs 6.19, 6.20 and 6.21.

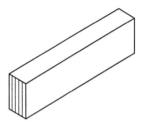

Fig. 6.19   Bar former.

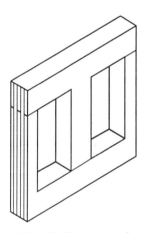

Fig. 6.20   Shell-type transformer.

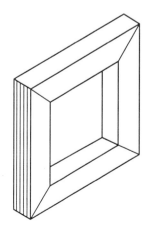

Fig. 6.21   Core-type transformer.

Other types are also manufactured for special purposes.

The steel formers are laminated; they are assembled from sheets coated on one side with a thin layer of insulating material, to reduce losses caused by eddy currents.

In the shell-type transformer both windings are located on the central arm of the former; this provides two magnetic circuits in parallel as shown in Fig. 6.22. With this type of winding a high volts per turn ratio can be achieved.

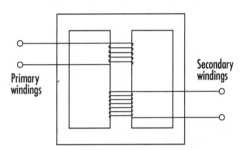

Fig. 6.22   Windings for a shell-type transformer.

In the core-type transformer the windings are wound on the outer arms as shown in Fig. 6.23.

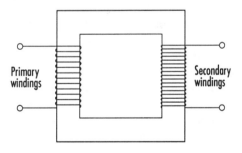

Fig. 6.23   Windings for a core-type transformer.

Many modern transformers have a preformed core of laminated strip made of grain-oriented cold-rolled silicon steel which allows a greater amount of freedom with transformer design. The British Standard graphical symbol for a transformer is shown in Fig. 6.24.

Fig. 6.24

When an a.c. supply is applied to the primary winding, a current will flow, producing a flux in the steel former. This flux links the primary and secondary coils, causing an e.m.f. to be induced in the latter.

A transformer therefore obtains an a.c. supply of a required voltage from an a.c. supply of another voltage both at the same frequency.

Transformers can be classified into three categories:

- power-frequency
- audio-frequency
- radio-frequency

The former of the power-frequency transformer is made from iron laminations; it operates at mains frequency and has a very high apparent power rating, up to several megavolt amperes. A very high operating efficiency ensures that the losses are kept low. The audio-frequency transformer, on the other hand, deals with relatively low power; it operates in the range of 15 Hz to 25 kHz and has a maximum apparent power rating of around 20 volt amperes (VA). The former is of shell-type construction and is comparatively small; it is made of silicon iron or granulated iron. Its main purpose is to give maximum flux linkage between the primary and secondary windings.

The coils of a radio-frequency transformer are wound on a core of ferrite, iron dust or simply air. With an iron core adjustments can be made to vary the coil inductance. Losses are kept low by using very thin laminations to minimize eddy currents.

The material used as a binder for the particles of iron also acts as an insulator which tends to break up the eddy current paths and reduce losses. The frequency range is wide, in the region of 1 kHz to 1 GHz.

Exercise 6.7

1. Sketch the three types of former that are in general use for transformer construction.

2. Explain why the former is often manufactured from laminated sheets rather than a solid block of iron.

3. Explain why in a shell-type transformer both windings are located on the centre limb.

4. Draw the British Standard graphical symbol for a transformer.

5. State the three general classifications of transformers.

6. List the frequency range of an audio-frequency transformer and a radio-frequency transformer.

7. Explain how the losses in a radio-frequency transformer are kept to a minimum.

6.13.2 E.m.f. equation for a transformer

The flux produced in the steel former of a transformer follows the waveform as shown in Fig. 6.25.

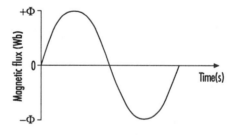

Fig. 6.25

In one half cycle the flux changes from $+\Phi$ to $-\Phi$, i.e. the flux change is 2Φ. The time taken for this flux change is $1/2f$ seconds. The average rate of change of flux is

$$\frac{\text{flux change}}{\text{time taken}} = \frac{2\Phi}{1/2f} = 4\Phi f \text{ weber/second}$$

Now $1\,V = 1\,Wb/s$. So the average e.m.f. induced is $4\Phi f$ volts.

If the number of turns on the coil is N then the average e.m.f. induced in the coil is $4\Phi fN$ volts.

Now r.m.s. value $= 1.11 \times$ average value. Therefore r.m.s. value of induced e.m.f. is

$$1.11 \times 4\Phi fN = 4.44\Phi fN$$

Consider a primary winding of N_1 turns and a secondary winding of N_2 turns. Then the respective r.m.s. values of induced e.m.f. are

$$E_1 = 4.44\Phi fN_1 \quad \text{and} \quad E_2 = 4.44\Phi fN_2$$

Example

1. The primary winding of a single-phase transformer is connected to a 240 V 50 Hz supply. If the maximum value of the core flux is 0.003 Wb and the secondary winding has 2000 turns calculate the secondary induced voltage.

$$\Phi = 0.003\,Wb \quad f = 50\,Hz \quad N_2 = 2000$$

Induced e.m.f. in secondary is given by

$$E_2 = 4.44\Phi fN_2 = 4.44 \times 0.003 \times 50 \times 2000$$

$$= 1332\,V$$

6.13.3 Voltage, current and turns ratio

We have established that a transformer has no moving parts and that it is the flux field created in the primary winding that links with the secondary winding to give a transformer its unique set of characteristics.

Because transformers are used with an a.c. supply we are unable to use the d.c. power formula of $P = VI$, so for a.c. we use $P = VI\cos\phi$ where $\cos\phi$ is the power factor. In transformers we are interested in the part of the formula using VI, so we give it a new name, apparent power (S). So $S = VI$. Transformers are given an apparent power rating measured in volt amperes or kilovolt amperes (kVA).

With this in mind we can now develop the relationship between the voltage, current and turns. In an ideal transformer:

$$\text{input energy} = \text{output energy}$$

Therefore:

$$\text{input power} = \text{output power}$$

or

$$V_1 I_1 \cos\phi_1 = V_2 I_2 \cos\phi_2$$

At full load $\cos\phi_1 = \cos\phi_2$. Therefore:

$$V_1 I_1 = V_2 I_2$$

so

$$\frac{I_1}{I_2} = \frac{V_2}{V_1} \qquad \text{(Eq. 1)}$$

In the ideal transformer, primary magnetomotive force = secondary magnetomotive force, i.e.

$$I_1 N_1 = I_2 N_2$$

and

$$\frac{I_1}{I_2} = \frac{N_2}{N_1} \qquad \text{(Eq. 2)}$$

Combining Eqs 1 and 2:

$$\frac{I_1}{I_2} = \frac{V_2}{V_1} = \frac{N_2}{N_1}$$

Examples

1. Calculate the full load primary and secondary currents of a single-phase transformer rated at 500/240 V and 15 kVA.
    Transformer rating $(VA) = V_2 I_2$. Therefore

$$I_2 = \frac{15 \times 1000}{240} = 62.5\,A$$

$I_1/I_2 = V_2/V_1$ so

$$I_1 = \frac{V_2 I_2}{V_1} = \frac{15 \times 1000}{500} = 30\,A$$

2. A 400/6600 V single-phase transformer has 2000 turns in the secondary winding.

Calculate the number of turns in the primary winding.

$$V_1 = 400\,\text{V} \quad V_2 = 6600\,\text{V}$$

Using $V_2/V_1 = N_2/N_1$

$$\frac{6600}{400} = \frac{2000}{N_1}$$

$$N_1 = \frac{2000 \times 400}{6600} = 121$$

3. A 5 kVA single-phase transformer has 900 turns in the primary winding and 150 turns in the secondary winding. When the primary winding is connected to 550 V calculate: (a) the secondary voltage; (b) the primary current, (c) the secondary current.

$$N_1 = 900 \quad N_2 = 150 \quad V_1 = 550\,\text{V}$$

(a) Using $V_2/V_1 = N_2/N_1$

$$V_2 = \frac{N_2 V_1}{N_1} = \frac{150 \times 550}{900} = 91.67\,\text{V}$$

(b) Using $5\,\text{kVA} = V_2 I_2 = 5 \times 1000$

$$I_2 = \frac{5000}{V_2} = \frac{5000}{91.67} = 54.54\,\text{A}$$

(c) Using $I_1/I_2 = N_2/N_1$

$$I_1 = \frac{N_2 I_2}{N_1} = \frac{150 \times 54.54}{900} = 9.09\,\text{A}$$

Exercise 6.8

1. Assuming a sinusoidal flux variation with time, derive an expression for the induced e.m.f. of a transformer, given the frequency, number of turns per winding, maximum flux density, and effective core cross-sectional area. A 3300/250 V 50 Hz, single-phase transformer is built on a core having an effective cross-sectional area of 125 cm^2 and 70 turns on the low-voltage winding. Calculate: (a) the value of the maximum flux density; (b) the number of turns on the high-voltage winding

2. Derive an expression for the r.m.s. value of the e.m.f. induced in a coil of N turns by a sinusoidally varying flux (maximum value Φ_m) of frequency f.

   A single-phase, filament transformer is required to supply 15 A at 6.3 V when operating from 200 V 50 Hz mains. The core area is 938 mm^2 and the maximum flux density is 0.867 T. Neglecting all losses and the effect of the magnetizing current, determine: (a) the supply current; (b) the number of turns on the primary winding; (c) the number of turns on the secondary winding.

3. Enumerate the losses which occur in a transformer when supplied from a constant-voltage, constant-frequency source. Explain how the losses vary with load.

   Show that the maximum efficiency occurs when the load is such that it makes the copper loss equal to the iron loss.

   The primary winding of a single-phase transformer is connected to a 230 V 50 Hz supply. The secondary winding has 1500 turns. If the maximum value of the core flux is 0.002 07 Wb, determine: (a) the number of turns on the primary winding; (b) the secondary induced voltage; (c) the net cross-sectional core area if the flux density has a maximum value of 0.465 T.

4. Make a neat diagram or sketch of a simple single-phase double-wound transformer, and with it explain the action of the transformer.

   Calculate the respective number of turns in each winding of such a transformer which has a step-down ratio of 3040 V to 240 V if the 'volts per turn' are 1.6.

5. A simple transformer has 1200 turns on the primary coil and 300 turns on the secondary coil. When the primary voltage is 240 V, determine the value of the secondary voltage.

6. Explain briefly the action of a transformer and show that the voltage ratio of the primary and secondary windings is the same as their turns ratio.

A 50 kVA, single-phase, 50 Hz, 500/100 V transformer has an effective core cross-section of 130 cm². Calculate: the number of turns, and the conductor cross-section of both the high-voltage and low-voltage windings. Assume a maximum flux density of 1.1 T and a current density of 200 A/cm².

7. (a) Draw a detailed sketch of a shell-type transformer, giving an explanation of each part. (b) The primary winding of a single-phase double-wound transformer has 600 turns and is supplied with 400 V. Calculate the number of secondary turns required to give a no-load output of 36 V.

8. A 10 kVA, single-phase transformer has a turns ratio of 300/23. The primary is connected to a 1500 V 60 Hz supply. Find the secondary volts on open circuit, and the approximate values of the currents in the two windings on full load. Find also the maximum value of the flux.

9. A step-down transformer with a turns ratio of 16:1 has a rated output of 150 VA at 15 V. On full load it supplies current to two soil-warming wires of identical length which are connected in parallel. The wire diameter is 1.5 mm and its resistivity is 16.7 $\mu\Omega$ mm. Assume unity power factor. Determine: (a) the primary voltage and the current taken by the transformer, ignoring transformer losses; (b) the length of each wire.

10. (a) Describe the construction, and explain the operation of a single-phase double-wound transformer, using sketches and diagrams to illustrate your answer. (b) Write down the formulae relating primary and secondary voltages, currents, and numbers of turns. (c) A single-phase 6600/440 V transformer has a $33\frac{1}{3}$% tapping on the secondary winding. What secondary voltages will be available?

11. Describe, with the aid of a sketch, the construction and principle of operation of a double-wound transformer.

    A single-phase step-down double-wound transformer has 1000 turns on the primary winding. If the primary voltage is 500 V, and the secondary voltage 50 V, calculate the number of turns on the secondary and also the primary current when the secondary current is 10 A.

12. The 'volts per turn' of a single-phase double-wound transformer are 1.9. If the transformer has a step-down ratio of 4250 V to 247 V, and the secondary current is 125 A, calculate the number of turns in each winding and the primary current.

6.13.4 Step-up-transformer

The name of this transformer gives a good clue to its use. If we consider 230 V a.c. on the primary side of the step-up transformer we will have on the secondary side a higher voltage. The input voltage has been stepped up. A general system is shown in Fig. 6.26.

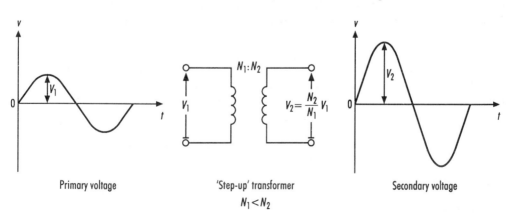

Primary voltage   'Step-up' transformer   Secondary voltage
                              $N_1 < N_2$

Fig. 6.26

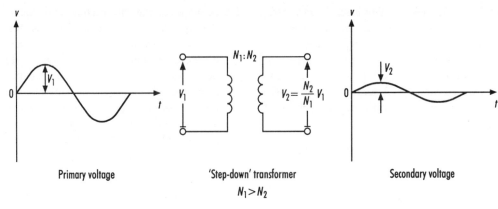

Primary voltage   'Step-down' transformer   Secondary voltage

$$N_1 > N_2$$

Fig. 6.27

A most important use of a step-up transformer is on the National Grid. The output from the power station of 22 or 33 kV is transformed up to 275 or 132 kV for the British National Grid or up to 400 kV for the National Supergrid. The reason for using such high voltages is that it is cheaper in transmission to use a high voltage and a low current than to use a high current and a low voltage. After the electrical energy has been transmitted it needs to be distributed to industrial outlets, commercial premises and domestic dwellings. For your household, this very large voltage, in a series of steps, is lowered to 230 V 50 Hz.

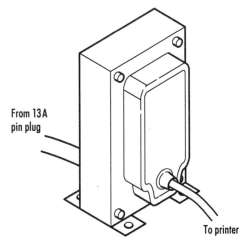

From 13 A
pin plug

To printer

Fig. 6.28

6.13.5 Step-down transformer

Again the name of this transformer gives a good clue to its use. If we consider 230 V a.c. on the primary side of the step-down transformer we will have on the secondary side a lower voltage. The input voltage has been stepped down. A general system is shown in Fig. 6.27.

Applications of this type of transformer are wide ranging. In my home the first one to be spotted is connected to my computer printer from which this text is originally printed. It was made in Mexico with a technical specification as follows:

| Input: | 240 V | 50 Hz | 90 mA | 23 VA |
| Output: | 30 V | | 400 mA | 12 W |

The physical size is 90 mm × 65 mm × 50 mm. An outline drawing is shown in Fig. 6.28.

6.13.6 Isolating transformer

This type of transformer is used to isolate any circuit or device from its power supply so its basic function is as a safety device. Applications for this type of transformer are in control circuits with features such as: low-loss laminations welded to prevent vibration and epoxy-painted against rust; a chrome electroplated zinc frame; a resin-impregnated bobbin with an isolating protective cover.

A typical specification is:

Primary: 230/240 V a.c.
Secondary: 12 V a.c.

For a rating of 300 VA the physical size is 125 mm × 155 mm × 125 mm with a weight of about 8 kg.

An important use of the isolating transformer is for outdoor work. In general, they are high quality, rugged and designed to ensure safe site outdoor working voltages for power tools and lighting. The transformers are often encased in bright yellow plastic. Electrical protection against overload is dealt with by using a re-settable thermal cut-out. When necessary the primary voltage and secondary voltage are both 230 V a.c.; however, when outside on site they are often 230 V a.c. primary, 110 V a.c. secondary. The rating is up to 2250 VA with dimensions of 185 mm × 185 mm × 250 mm high and weighing nearly 20 kg.

For indoor work, an isolation transformer is used in the music industry, for example, to isolate an electric guitar from a 230 V a.c. supply.

6.13.7 Auto-transformer

The British Standard graphical symbol is shown in Fig. 6.29. The auto-transformer has one coil and the input voltage (V_i) and output voltage (V_0) are shown on the diagram.

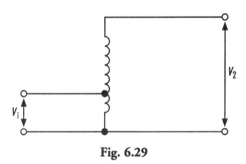

Fig. 6.29

Two advantages of this transformer are that it has only one coil which results in a cost saving, and it gives a higher operating efficiency compared with the step-up or step-down transformer. In addition the volume and weight of the transformer are less. The disadvantages are: (a) a break in the wire of the secondary winding will allow the primary voltage to be across the load, which could result in the load being damaged; and (b) if the secondary winding has a short-circuit fault the current will increase and cause damage to the load.

Some manufacturers enclose the transformer in an insulated flame-retardant plastic container.

Another feature is that in some cases the output is protected by a high breaking capacity (HBC) fuse.

6.13.8 Variable auto-transformer

With a variable auto-transformer, instead of having just one output, a range of values can be obtained by rotating a knob located on top of the transformer. A typical variable auto-transformer is shown in outline in Fig. 6.30. A typical rating for this transformer is 1 kVA with physical dimensions of 150 mm diameter 100 mm height. This type of transformer is used in a variety of applications, such as motor speed control, voltage regulation, laboratory power supplies, test equipment, etc.

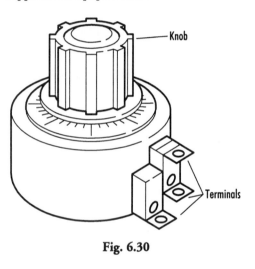

Fig. 6.30

6.13.9 The transformer as a matching device

A transformer may be used as a resistance matching device between a load having a relative low resistance and an amplifier having a relatively high resistance. The purpose of the transformer is to make the load resistance, as seen from the output terminals of the amplifier, appear to have the same value as the output resistance of the amplifier. This ensures maximum power transfer from the amplifier to the load.

Consider the circuit diagram shown in Fig. 6.31.

Fig. 6.31

The input resistance is given by

$$R_1 = \frac{V_1}{I_1} \quad \text{and} \quad V_1 = I_1 R_1$$

The output resistance is given by

$$R_2 = \frac{V_2}{I_2} \quad \text{and} \quad V_2 = I_2 R_2$$

From the voltage and turns ratio formula:

$$\frac{V_2}{V_1} = \frac{N_2}{N_1}$$

Since

$$\frac{I_2 R_2}{I_1 R_1} = \frac{N_2}{N_1}$$

$$R_2 = \frac{I_1 R_1 N_2}{I_2 N_1}$$

But

$$\frac{I_1}{I_2} = \frac{N_2}{N_1}$$

so, by substitution:

$$R_2 = \frac{N_2 R_1 N_2}{N_1 N_1} = R_1 \left(\frac{N_2}{N_1}\right)^2$$

and

$$R_1 = R_2 \left(\frac{N_1}{N_2}\right)^2$$

Examples——————

1. Calculate the value of a load resistor to give maximum transfer of power when connected to a power amplifier of output resistance $240\,\Omega$ through a transformer which has 300 turns on the primary winding and 100 turns on the secondary winding.

$$R_1 = 240\,\Omega \quad N_1 = 300 \quad N_2 = 100$$

Using $R_2 = R_1(N_2/N_1)^2$

$$R_2 = 240 \times \left(\frac{100}{300}\right)^2$$

$$= \frac{240}{9} = 26.67\,\Omega$$

2. An amplifier has an output resistance of $62.5\,\Omega$ and is to be connected to an external circuit of input resistance $1000\,\Omega$. For maximum power transfer calculate: (a) the transformer turns ratio; (b) the number of turns on the secondary winding if the primary winding has 100 turns.

$$R_1 = 62.5\,\Omega \quad R_2 = 1000\,\Omega \quad N = 100$$

From $R_2 = R_1(N_2/N_1)^2$

(a) $\dfrac{N_1}{N_2} = \sqrt{\left(\dfrac{R_1}{R_2}\right)} = \sqrt{\left(\dfrac{62.5}{1000}\right)}$

$$= \sqrt{0.0625} = 0.25$$

(b) $N_2 = \dfrac{N_1}{0.25} = \dfrac{100}{0.25} = 400$

3. A 40 V generator of resistance $75\,\Omega$ is to be matched to a $30\,\Omega$ resistor using a transformer. Calculate the transformer turns ratio and the power transferred to the load. Determine also the power which would be transferred if a direct connection were made from the generator to the load without using a transformer.

Comment upon the different values of power obtained in each case.

$$R_1 = 75\,\Omega \quad R_2 = 30\,\Omega \quad V_1 = 40\,V$$

Case 1: transformer in circuit

Using $N_1/N_2 = \sqrt{(R_1/R_2)}$

$$\frac{N_1}{N_2} = \sqrt{\frac{75}{30}} = \sqrt{2.5} = 1.58$$

For transfer of maximum power $R_1 = R_2$. The current flow is given by

$$I = \frac{V_1}{R_1 + R_2} = \frac{V_1}{R_1 + R_1} = \frac{V_1}{2R_1}$$

and the power by

$$P_2 = I^2 R_2 = I^2 R_1 = \left(\frac{V_1}{2R_1}\right)^2 R_1 = \frac{V_1^2}{4R_1}$$

Therefore

$$P_2 = \frac{40^2}{4 \times 75} = \frac{1600}{300} = 5.33\,\text{W}$$

Case 2: circuit without a transformer

Using $I = V_1/(R_1 + R_2)$

$$I = \frac{40}{75 + 30} = \frac{40}{105} = 0.381\,\text{A}$$

Using $P_2 = I^2 R_2$

$$P_2 = 0.381^2 \times 30 = 0.1452 \times 30 = 4.36\,\text{W}$$

The power transfer with the transformer connected is 5.33 watts and without it is 4.36 watts. This shows that more power (in fact maximum power) is transferred when the transformer is employed.

Exercise 6.9

1. Explain, with the aid of a sketch, the construction and action of: (a) a step-up transformer; (b) a step-down transformer.

2. Give examples of the use of a step-up and a step-down transformer.

3. What is the basic function of an isolating transformer? Give examples of its use indoors and on an outdoor site.

4. Explain, with the aid of sketches, the construction and action of an auto-transformer. State at least two advantages and two disadvantages of an auto-transformer as compared to a step-down transformer.

5. What is the difference between an auto-transformer and a variable auto-transformer?

6. Explain how a transformer can be used as a matching device.

7. A transformer has 375 turns on the primary winding and 75 turns on the secondary winding. Calculate the value of a load resistor to give maximum transfer of power when connected to a power amplifier of output resistance $150\,\Omega$.

8. A transformer has a primary winding of 200 turns. An amplifier has an output resistance of $75\,\Omega$ and is to be connected to an external circuit of input resistance $1200\,\Omega$. For maximum power transfer calculate: (a) the transformer turns ratio; (b) the number of turns on the secondary winding.

9. A 50 V generator of resistance $90\,\Omega$ is to be matched to a $22\,\Omega$ resistor using a transformer. Calculate the transformer turns ratio and the power transferred to the load.

10. A 60 V generator of resistance $68\,\Omega$ is to be matched to a $24\,\Omega$ resistor using a transformer. Calculate: (a) transformer turns ratio; (b) power transferred to the load; (c) power transferred to the load without using the transformer. Comment upon the difference in the values obtained in parts (b) and (c) of the question.

6.14 SWITCHES

In its simplest form, a switch is a device that opens or closes a circuit. When you look in a catalogue you will find a wide range of switches. Some will be considered in this chapter.

6.14.1 Single-pole single-throw switch

The British Standard graphical symbol for this switch is shown in Fig. 6.32. The term 'pole' means the number of separate circuits the switch can make at the same time, in this case one. The term 'throw' means the number of positions to which each pole can be switched, in this case one.

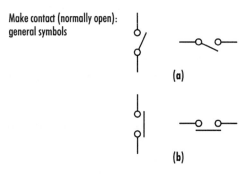

Make contact (normally open):
general symbols

(a)

(b)

Fig. 6.32

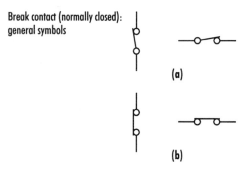

Break contact (normally closed):
general symbols

(a)

(b)

Fig. 6.33

As shown they can be drawn either vertically or horizontally. The switch contacts can move it in only one direction to open or close it. In diagram (b) the switch is called a push-button switch for fairly obvious reasons – it needs to be pushed to close it. As shown in both diagrams the make contact is normally open. The alternative to this is the break contact, which is normally closed as shown in Fig. 6.33. To keep the switch contacts in position a spring mechanism is used; this is called latching.

The circuit diagram shown in Fig. 6.34 is an application of such a switch. The circuit is that of a d.c. output power supply; it could also be used as a battery charger. It is interesting to note that this circuit uses most of the components that have been dealt with so far. When the switch is closed the neon lamp acts as an indicator to show that the system is ON. The letter E is there to remind you that in industrial and domestic circumstances the system should be earthed. This earth conductor is more correctly now called the circuit protective conductor (CPC) according to the latest wiring regulations.

6.14.2 Single-pole double-throw switch

The British Standard graphical symbol for a single-pole double-throw switch is shown in Fig. 6.35. This type of switch has many uses. A popular one is in the home for use at the bottom and top of the stairs. You can switch the

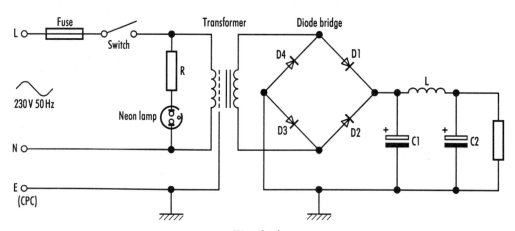

Fig. 6.34

light on downstairs, climb the stairs and switch off the light. Anyone wishing to climb the stairs later can still switch the light on from the bottom of the stairs and again climb the stairs and switch it off. A circuit diagram for this type of application is shown in Fig. 6.36.

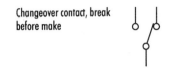

Changeover contact, break before make

Fig. 6.35

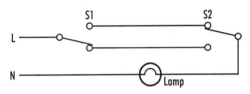

Fig. 6.36

A very useful use of this type of switch is in very long buildings. You can switch the lights on at one end of the building and switch them off at the far end of the building without having to go back to your starting point.

Later on in this book you will carry out an investigation using this circuit to find out what happens to the lamp when the switches are placed in various positions. In the position shown the lamp would be OFF.

6.14.3 Double-pole single-throw switch

The British Standard graphical symbol for this switch is shown in Fig. 6.37. This type of switch is used when a break is needed to be made in the live (L) and the neutral (N) conductor. For example, it could have been used in the power supply of Fig. 6.34.

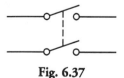

Fig. 6.37

6.14.4 Rotary switch

The British Standard graphical symbol is shown in Fig. 6.38 for a single-pole 12-position switch S1, one position being the off position, i.e. with no contact.

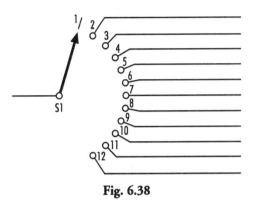

Fig. 6.38

With this type of switch it is usual to have a tag location drawing as shown in Fig. 6.39. In the tag location drawing the centre number indicates the number of positions, including the OFF position. The numbers shown on the outer circle are the tag numbers. A drawing of a rotary switch is shown in Fig. 6.40.

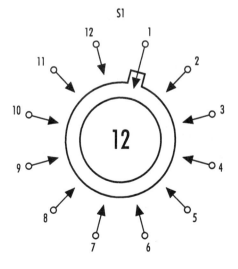

Fig. 6.39

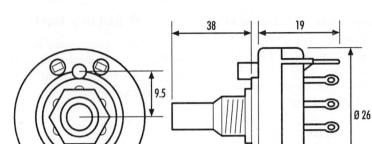

Fig. 6.40

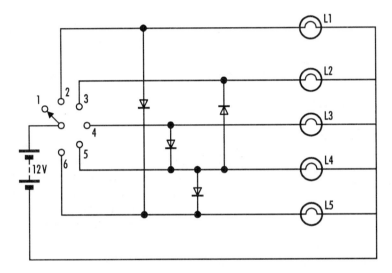

Fig. 6.41

Most rotary switches are of high quality and are moulded in glass-filled nylon. A typical specification is:

Contact rating	150 mA 250 V a.c.
Initial contact resistance	20 MΩ
Electrical life	15 000 full rotation cycles
Temperature range	−20 to +65°C

A circuit diagram shown in Fig. 6.41 shows five lamps connected in conjunction with four diodes using a single-pole six-position switch.

6.14.5 Reed switch

Reed switches are hermetically sealed and precisely made for safe and reliable use under critical conditions. A good quality reed switch is not usually affected by any atmospheric conditions, temperature or pressure. The construction of two types of reed switch is shown in Fig. 6.42.

The reed switch is operated by placing a magnet close to the glass container. The reeds become magnetized, attract each other, touch and complete a circuit. The alternative to using a magnet is to use a current-carrying coil which, when placed near or wrapped round the glass container, acts in the same way as before. When the magnet is removed or the current is switched off, the reeds will come apart, i.e. in the OFF position. The reed switch is useful when we need a small current in one circuit to control another circuit.

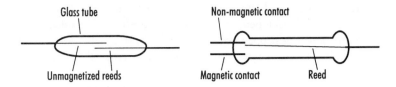

Fig. 6.42

6.14.6 DIL switch

DIL stands for 'dual-in-line'. This type of switch has applications in computing, communications equipment, remote control systems, security equipment, etc. It is usually rectangular in shape and comprises a number of single-pole single-throw switches. A typical outline drawing is shown in Fig. 6.43.

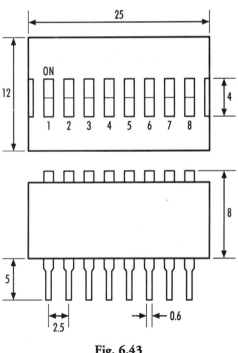

Fig. 6.43

The contacts are often gold-plated phosphor bronze, encapsulated in a plastic body that is flame-retardant with a sealed base to prevent flux contamination. The ON/OFF status is clearly identified on the switch as shown in the diagram.

6.14.7 'Emergency stop' push-button switch

This is probably the most important switch of them all. If you are in a laboratory or workshop and an accident happens it is important that the electrical supply is cut off from whatever is causing the accident. Whatever type of room you are in, the first lesson should include the identification of the position of the emergency stop buttons. The switch is usually painted in RED and has a mushroom-shaped cap with arrows. Clearly mounted above the switch is a plate that states 'Emergency STOP'. The switch stays in the OFF position until activated by striking it and can only be released when twisted in the direction of the arrows. The physical size of the red cap is about 50 mm diameter. A typical switch is shown in Fig. 6.44.

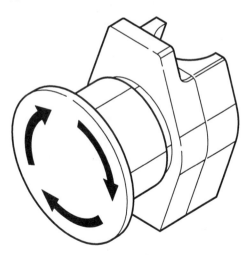

Fig. 6.44

6.14.8 Keyboard switch

Most of the switches we have mentioned can create a problem when used in electronic circuits. Instead of having a clean touch the switch contacts tend to bounce. This bounce can repeat several times whatever signal is being sent whereas we really only want it to happen once. In sophisticated electronic circuits we can use an electronic de-bounce circuit. Sometimes the alternative is to use a 'touch' switch. These are used on computer keyboards, calculators, etc. Keyboard switches can either have round or square tops and are primarily designed for digital applications. A typical specification is:

Maximum load	100 V a.c. 10 mA
Insulation resistance	Greater than 1 MΩ
Contact resistance	Less than 10 mΩ
Life	More than 250 000 operations
Operating force	1.3 N
Travel	0.8 mm
Operating temperature	−20 to +65°C

When using a keyboard switch we simply touch the pad, the electronic circuit reacts and there is no bounce.

6.15 RELAYS

A relay can be considered as another form of switch. The British Standard symbol for a single-element relay and the alternative symbol are shown in Fig. 6.45.

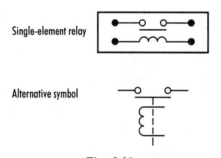

Fig. 6.45

A bell circuit is shown in Fig. 6.46. When the bell push is pressed the bell should ring. If the cable is too long the sound of the bell will be very low. A relay can be used in a circuit to overcome this problem as shown in Fig. 6.47. In the original circuit the voltage drop along the cable reduces the supply voltage to the bell by a

considerable amount. But by placing the relay close to the bell only a small current is needed to create the magnetic field to close the relay contacts and the bell sounds normally.

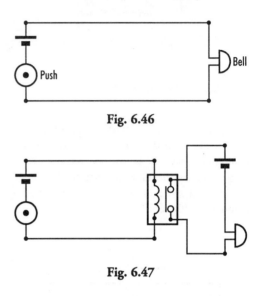

Fig. 6.46

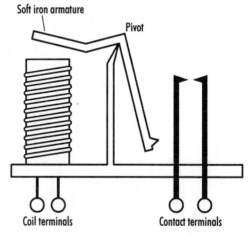

Fig. 6.47

There are several different relay designs. A typical construction is shown in Fig. 6.48.

Fig. 6.48

In the bell circuit, in Fig. 6.47, when the push is pressed, a current flows through the relay coil, the coil creates a magnetic field and the soft iron armature is attracted to it. The other end of the armature hits one of the contact terminals and forces it against the other contact, thus completing the circuit and causing the bell to ring.

Description	Symbol
Relay coil: general symbol	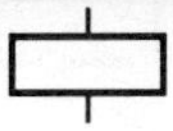
Example: relay coil with 1300 Ω winding	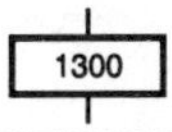 1300
Coil of a slow-releasing relay	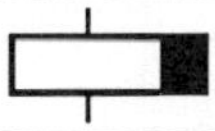
Coil of a very slow-releasing relay	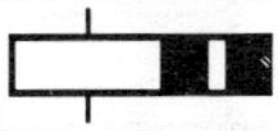
Coil of a slow-operating relay	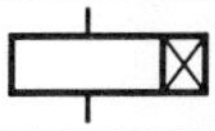
Coil of a very slow-operating relay	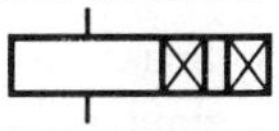
Coil of a slow-operating relay and a slow-releasing relay	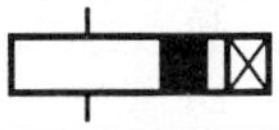
Coil of an a.c. relay	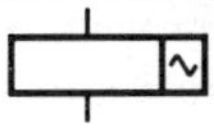
Note: the frequency of the operating current may be indicated if desired. Example: coil of a relay operated by a current of 50 Hz	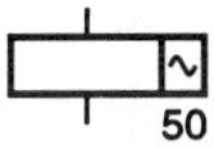 50 Hz
Coil of a relay unaffected by a.c.	
Relay coil offering high impedance to speech currents	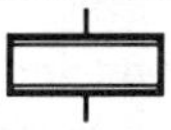
Coil of a high-speed relay	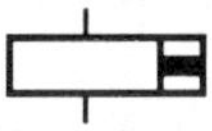

Fig. 6.49

Description	Symbol
Make contact unit	
Break contact unit	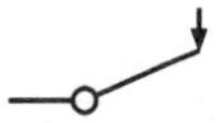
Changeover (break before make) contact unit	
Changeover (make before break) contact unit	
Changeover (both sides stable) contact unit	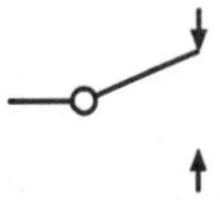
Two-way contact unit with neutral position	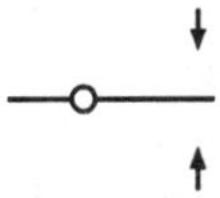
Contact unit with two makes, making in succession	
Contact unit with two breaks, breaking in succession	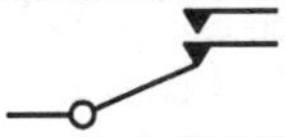
Two-way contact unit with mechanically resonant blade	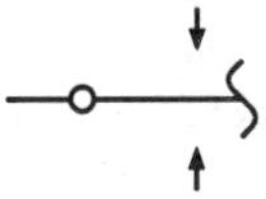

Fig. 6.50

In relays there are two types of symbol that need to be considered. Figure 6.49 shows a selection of coil types, i.e. the device that causes the switch contacts to operate.

Figure 6.50 shows a selection of the relay contact units. When using a relay it is important

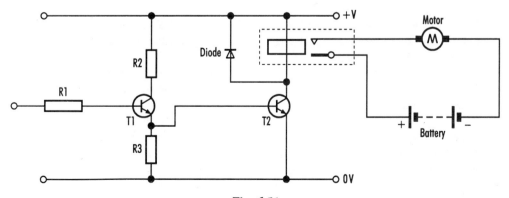

Fig. 6.51

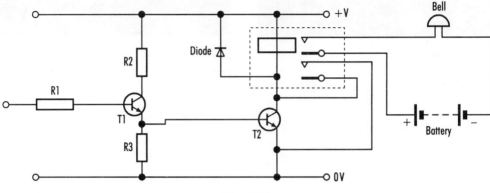

Fig. 6.52

to consider carefully the data sheets provided by the manufacturer so that the correct connections are chosen for the different types of relay available. Figure 6.51 on page 111 shows part of an electronics circuit.

A diode is included to protect the circuit because the relay coil acts as an induction coil which can give a large reverse voltage. The diode short circuits this voltage and saves the circuit from being damaged.

When the relay is activated it can then run any number of items, such as a motor, a bell, a buzzer, a counter, etc. A motor is shown connected.

Figure 6.52 shows part of an electronics circuit which is using a second set of relay contacts to act as a latch. The bell will continue ringing after the initial current flowing through the relay coil has stopped. This is part of a circuit that could be used as a detector alarm.

Exercise 6.10

1. Explain what is meant by the term switch.

2. Draw two sets of BS graphical symbols for a single-pole single-throw switch. Explain what is meant by the terms pole and throw.

3. Why is it necessary to have a spring mechanism in a single-pole switch?

4. Draw a labelled circuit diagram of a power supply using a single-pole single-throw switch. Explain, with another diagram, how the single-pole switch can be replaced by a double-pole single-throw switch.

5. State an application for a single-pole double-throw switch and then draw a circuit diagram of your stated application.

6. Explain, with the aid of a circuit diagram, how a rotary switch can be used.

7. Explain, with the aid of sketches, the construction and action of a reed switch.

8. Explain what is meant by the term DIL switch and state at least two applications for such a switch.

9. Explain, with the aid of a sketch, the action and purpose of an 'Emergency stop' push-button switch.

10. What is meant by the term bounce?

11. State the main reason for having keyboard switches and give at least two examples of such a switch.

12. Explain what is meant by the term relay using a British Standard graphical symbol.

13. Describe, with the aid of a sketch, the physical construction and action of a relay.

14. Explain, with the aid of a diagram, how a relay can be used to enhance the performance of a bell circuit.

15. Draw BS graphical symbols for eight coil types and eight relay contact units.

16. State why a diode is always used in conjunction with a relay.

17. Draw part of a circuit diagram that shows a counter being activated by a relay circuit.

18. Draw part of a circuit diagram using a relay that has been latched. State one use of such a circuit.

19. Look in a manufacturer's catalogue and make a list of switches and applications not mentioned in this chapter.

Answers

Exercise 6.1
1. 5 N
2. 0.4952 T
3. 0.0675 N, 0.00027 Nm
4. 10.8 N
5. 8.55 N, 1.7 Nm
6. (a) 0.002 N; (b) 0.000 03 Nm
7. 90 N

Exercise 6.2
1. (a) 2.5 N; (b) 1.5 V; (c) 30 W
2. (b) 1.556 mV
3. 0.2 V
4. 3.142 V
5. (a) 10 m/s; (b) 0.8 N; (c) 0.48 J
6. (b) 1.5 T, 5.4 N
7. 9.6 V, 12 N, 480 W
8. 31.4 V
9. 0.745 V

Exercise 6.3
1. 0.25 H, 75 V
2. 0.09 H, 180 V, 4.5 J
3. 0.15 H, 30 V
4. (a) 10 kV; (b) 700 V
5. 0.2 H, 40 V
6. 20 V
7. (c) 5 mV
8. (b) 20 kV
9. 15 H, 300 V
10. (a) 18×10^8 V, 9×10^6 V
11. 466.7 V, 5.6 H

Exercise 6.4
1. (a) 1600 J; (b) 40 A
2. (a) 1 J; (b) 6 Ω
3. (a) 1.5 Ω; (b) 8 A
4. 1800 μJ
5. 212.1 V, 75.44 mH
6. 7 mJ
7. (b) 15 H, 1.2 J, 60 V
8. 2.5 H, 20 J
9. 2230, 0.02 J

Exercise 6.5
1. 3.142 Ω
2. 0.1591 H
3. (a) 6.365 A; (b) 0.3182 A
4. 40 Ω, 0.1273 H
5. 0.0764 H, 0.009 548 H
   251.4 V, 10 V
   2.00 A
   15.91 Hz
   125.7 Ω, 12 Ω
6. (a) 17.14 A; (b) 9.231 A

Exercise 6.6
1. 0.2 s, 3.94 A, 10 A, 0.139 s
2. (a) 2.5 A/s; (b) 0.1426 A; (c) 0.055 44 s
3. (a) 20 A/s; (b) 0.1 s; (c) 1.5 A
4. (a) 0.002 s; (b) 0.6 H; (c) 0.366 A; (d) 183 A/s
5. 4.68 ms, 0.4 J
6. 277.5 ms
7. (a) 50 A/s; (b) 0.5 A
8. (a) 50 ms; (b) 50 V; (c) 20 A/s; (d) 20 J

Exercise 6.8
1. (a) 1.289 T; (b) 924
2. (a) 0.472 A; (b) 1117.8; (c) 35.2
3. (a) 505; (b) 683 V; (c) 44.5 cm^2
4. 1900, 150
5. 60 V
6. 159, 0.5 cm^2, 32, 2.5 cm^2
7. (b) 24
8. 115 V, 6.67 A, 87 A, 18.75 mWb
9. (a) 240 V; 0.625 A; (b) 317.5 m
10. (c) 440 V, 146.7 V, 293.3 V
11. 100, 1 A
12. 2237, 130, 7.264 A

Exercise 6.9
1. See 6.13.4, 6.13.5
2. See 6.13.4, 6.13.5

3. See 6.13.6

4. See 6.13.7

5. See 6.13.7, 6.13.8

6. See 6.13.9

7. $6\,\Omega$

8. (a) 0.25; (b) 800

9. 2.0226, 6.94 W

10. (a) 1.683; (b) 13.24 W; (c) 10.21 W

Exercise 6.10

1. See 6.14

2. See 6.14.1

3. See 6.14.1

4. See 6.14.1, 6.14.3 Fig. 6.34

5. See 6.14.2

6. See 6.14.4

7. See 6.14.5

8. See 6.14.6

9. See 6.14.7

10. See 6.14.8

11. See 6.14.8

12. See 6.15, Fig. 6.45

13. See 6.15, Fig. 6.48

14. See 6.15, Figs 6.46, 6.47

15. See 6.15, Figs 6.49, 6.50

16. See 6.15

17. See 6.15, Fig. 6.51

18. See 6.15, Fig. 6.52

7

COMPONENTS 4

7.1 DIODE

This is an electronic device that has two electrodes. The main property of a diode is that it will only allow current to pass in one direction.

There are several different types of diode but they are most commonly used as rectifiers. It is vital that they are connected into a circuit correctly otherwise damage will take place. Usually the anode goes to the positive terminal of the supply and the cathode end, having a marked band on it, goes to the negative terminal. Figure 7.1 shows the diode connected correctly, in what is called forward bias; current flows and the lamp lights.

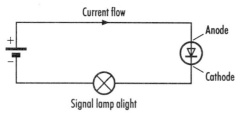

Fig. 7.1

With forward bias the diode has a very low resistance and this is why the lamp lights. If the diode had been connected the other way round, then it would have reverse bias, a high resistance, and the current would not flow.

7.2 FORWARD AND REVERSE BIAS

As illustrated in Fig. 7.2 when an e.m.f. is applied to a p–n junction the junction is said to be 'forward-biased'.

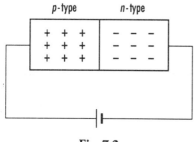

Fig. 7.2

Because like charges repel, positive holes are repelled from the positive terminal of the cell; similarly the negative electrons are repelled from the negative terminal. The free holes and electrons moving through the material can penetrate into the depletion layer, and through it, to recombine. For each electron that is lost due to recombination, another electron enters the n-type material from the negative terminal of the cell. Likewise, each time a hole recombines with an electron at the junction, another hole forms near the positive terminal of the cell when an electron breaks its bond and enters the positive terminal. Since electrons are entering the n-type material and leaving at the positive end, an electron flow is established in the external circuit. With the cell connected in this 'forward-bias' manner there is a current in the p-type material carried by the holes and a current in the n-type material carried by the electrons. Forward bias results in a flow of current in the external circuit.

Reverse bias is when the cell connections are reversed as illustrated in Fig. 7.3.

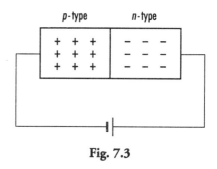

Fig. 7.3

The electrons, being negative charges, are attracted to the positive terminal of the cell since unlike charges attract each other. For the same reason, the holes, being positive charges, are attracted to the negative terminal of the cell. With holes and electrons being drawn further from the junction, the depletion layer becomes wider. Since there are hardly any mobile charge carriers on either side of the junction, very little current can flow. Typical voltage–current curves for forward- and reverse-bias conditions are illustrated in Fig. 7.4.

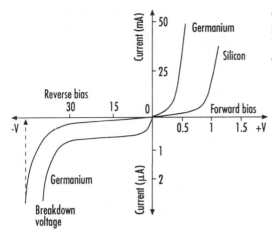

Fig. 7.4

circuit is shown in Fig. 7.6. The output wave-form is shown in Fig. 7.7. The diode in circuit Fig. 7.6 conducts when the anode is positive with respect to the cathode.

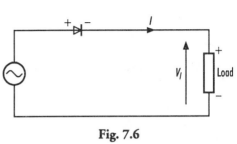

Fig. 7.6

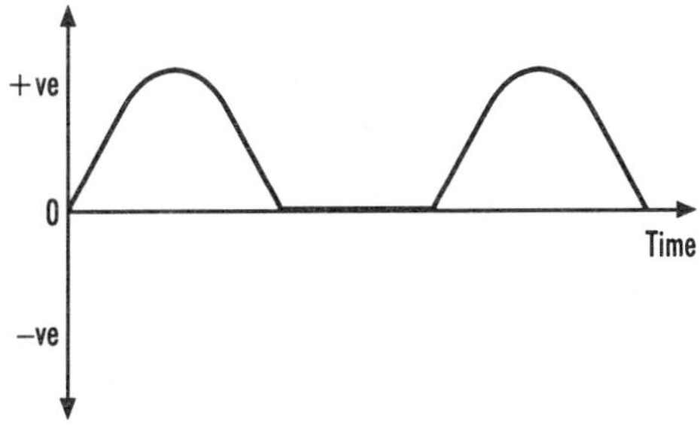

When a particular value of reverse-bias voltage is exceeded a relatively large current flows. This value of voltage is identified in Fig. 7.4 as the breakdown voltage.

7.3 CURRENT AND ELECTRON FLOW

The circuit illustrated in Fig. 7.5 indicates the direction of conventional current flow and elec-tron flow. Notice that electron flow is in the opposite direction to conventional current flow; it is important to indicate which of the two is being used.

Finally, it is important to note that the phrase 'majority carriers' refers to electrons in *n*-type material or holes in *p*-type material. The less concentrated carriers are then referred to as 'minority carriers' – that is, holes in *n*-type material and electrons in *p*-type material.

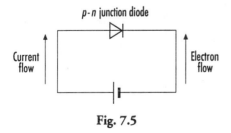

Fig. 7.5

7.4 RECTIFICATION

Rectification means changing an a.c. to a fluc-tuating current. A simple half-wave rectifier

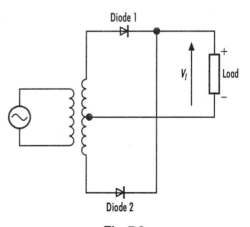

Fig. 7.7

Full-wave rectification can be obtained by using two diodes in a circuit as shown in Fig. 7.8. Diodes 1 and 2 conduct in alternative half cycles. The output waveform is shown in Fig. 7.9.

Fig. 7.8

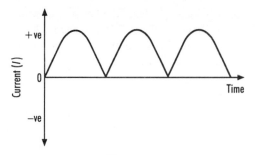

Fig. 7.9

and 7.12. The diodes are arranged in the bridge circuit so that in the circuit of Fig. 7.11 the current flows through diodes 1 and 3 and in the circuit of Fig. 7.12 the current flows through diodes 2 and 4. By careful inspection it can be seen that the current flowing through the load resistor is unidirectional.

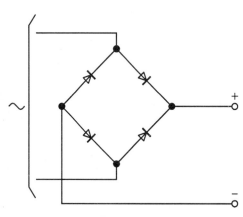

Fig. 7.10

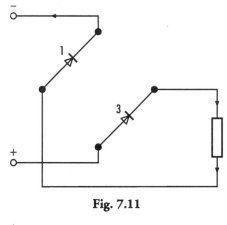

Fig. 7.11

Fig. 7.12

This circuit is not often used because of the weight and relatively high cost of the transformer. A bridge circuit, as shown in Fig. 7.10, is used which does not need a centre-tapped transformer. The shape of the output waveform is the same as that in Fig. 7.9.

It is easier to understand the output waveform in Fig. 7.9 if the circuit in Fig. 7.10 is divided into two sections as shown in Figs 7.11

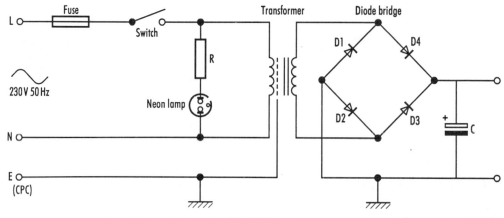

Fig. 7.13

7.5 POWER SUPPLY

Almost all electronic circuits need a d.c. power supply. Batteries can be used but usually it is more convenient to use power supply units operated from the 230 V 50 Hz a.c. supply. The complete circuit for a d.c. supply is shown in Fig. 7.13 at the bottom of the facing page.

We have used a step-down transformer, a rectifier bridge circuit and a capacitor. The electrolytic capacitor is used to smooth out the varying d.c. output seen in Fig. 7.9 to give a steady d.c. The smoothed d.c. output is shown in Fig. 7.14 as a full line. The dotted lines show the unsmoothed output voltages for each part of the bridge circuit.

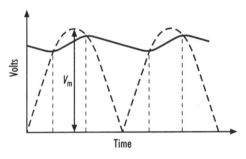

Fig. 7.14

More sophisticated circuits for smoothing are shown in Fig. 7.15. In Fig. 7.15(a) capacitor C1 is the reservoir capacitor referred to before as the smoothing capacitor, capacitor C2 is the filter capacitor and the resistor R is a filter and limiting resistor. In Fig. 7.15(b) the limiting resistor is replaced by an inductor because the potential difference across the resistor may become exces-

sive and cause damage. The inductive reactance will tend to limit the current. The capacitors in this circuit are the same as in the previous one.

7.6 ZENER DIODE

The BS graphical symbol and a characteristic graph for a zener diode are shown in Fig. 7.16 on page 120. In a semiconductor diode if you exceed the peak inverse voltage the diode will be damaged. A zener diode has the property that it will always break down at its rated reverse breakdown voltage and not be damaged. This property is very useful in many applications, one of which is the stabilized power supply, because it always needs a constant voltage output.

A current-limiting resistor is always used so that the zener diode is not damaged by an overload of current. The circuit diagram shown in Fig. 7.17 on page 120 would be placed after the smoothing circuit previously considered.

7.7 LIGHT-EMITTING DIODE

The British Standard graphical symbol for an LED is shown in Fig. 7.18 on page 120 with a current-limiting resistor. The LED will emit light when it is forward biased as shown in the circuit diagram. The colour of light emitted depends upon the manufacturer's specification. Popular colours are red and green but others are available. An LED will not emit light when it is reverse biased and in this mode can be easily damaged. A general use for an LED is as a signal lamp often indicating whether the equipment is ON or OFF.

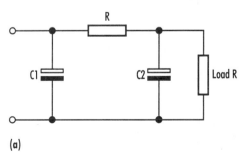

(a)

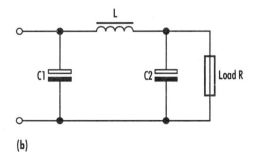

(b)

Fig. 7.15

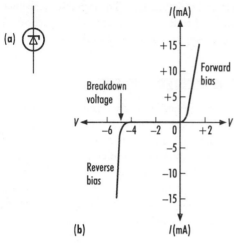

Fig. 7.16

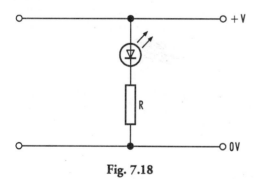

Fig. 7.17

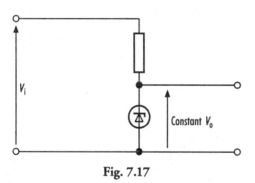

Fig. 7.18

Example

1. An LED is rated at 2 V 30 mA and is connected as shown in Fig. 7.18 to a 15 V supply. Determine: (a) the potential difference across the resistor; (b) the current flowing through the resistor; (c) the calculated value of the resistor; (d) the preferred value of the resistor from the E12 series.

(a) Potential difference across the resistor $= V_R$

$$V_R = 15 - 2 = 13 \text{ V}$$

(b) Because the LED is rated at 30 mA this must be the current through the resistor because they are connected in series. Therefore

$$I = 0.03 \text{ A}$$

(c) The value of resistor R is given by

$$R = \frac{V_R}{I} = \frac{13}{0.03} = 433.3 \, \Omega$$

(d) From the E12 series of resistors, the value above $433.3 \, \Omega$ is $470 \, \Omega$ and the value below $433.3 \, \Omega$ is $390 \, \Omega$. In practice we would use the $470 \, \Omega$ resistor because it would take a smaller current than the other resistor thus avoiding any likely circuit damage.

Exercise 7.1

1. Explain what is meant by the term rectification.

2. Explain, with the aid of circuit diagrams, the construction and action of the following types of rectifier: (a) half-wave; (b) centre-tap transformer full-wave; (c) bridge circuit full-wave.

3. Explain what is meant by the term smoothing circuit.

4. Draw a BS graphical symbol and a characteristic graph for a zener diode.

5. Explain why a resistor is always used in conjunction with a zener diode.

6. Explain, with the aid of a sketch, the construction and action of an LED.

7. Draw a labelled circuit diagram showing an LED connected in series with a resistor connected to a supply. Explain the purpose of the resistor.

8. An LED is rated at 2 V 10 mA and is connected in series to a resistor. If the supply is 6 V determine: (a) the current flowing through the resistor; (b) the potential difference across the resistor; (c) the calculated value of the resistor; (d) the preferred value of the resistor from the E24 series.

9. Draw on the same axes, choosing suitable scales, typical characteristics for silicon, germanium and zener diodes.

10. State one use of a zener diode and explain its main characteristic. State the voltage conditions and bias conditions for the best use of a zener diode. Explain why a zener diode must always be connected in series with a resistor.

11. Calculate the maximum current a zener diode can carry without damage if its breakdown voltage is 10 V and its maximum power rating is 2.5 W.

12. A zener diode rated at 5 W with a breakdown voltage of 6 V to be used to supply a constant voltage of 6 V from a 12 V battery. Calculate: (a) maximum diode current; (b) the value of a series resistor needed to avoid damage in the circuit.

13. From the given data plot the reverse-bias characteristics for a zener diode.

Current (mA)	0	200	400	600	800
Voltage (V)	9.8	10.1	10.2	10.4	10.8

14. An LED rated at 2 V is connected in series with a 470 Ω resistor. The supply is 12 V. Draw a circuit diagram and calculate: (a) the potential difference across the resistor; (b) the current flowing through the resistor; (c) the current rating of the LED.

15. In which of the following circuits shown in the figure would the LED light? Give a reason for each of your choices. Explain why the 560 Ω resistor is needed in each of the circuits.

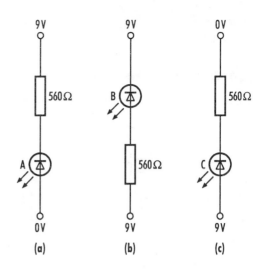

(a)   (b)   (c)

7.8 JUNCTION TRANSISTOR

A transistor is a combination of two *p-n* junctions which give either a *p–n–p* transistor or an *n–p–n* transistor as shown in Fig. 7.19.

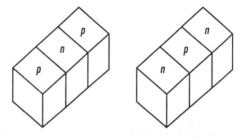

Fig. 7.19

The three regions are known as the emitter, the base and the collector and can be identified according to the graphical symbols shown in Fig. 7.20.

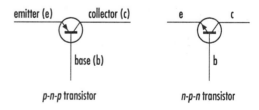

Fig. 7.20

The arrows on the British Standard graphical symbols point in the direction of conventional current flow, i.e. hole flow. The principle of operation is the same for both types of transistor, the difference being the polarity of the d.c. supply.

7.9 TRANSISTOR ACTION

The basic circuit for a $p{-}n{-}p$ transistor is shown in Fig. 7.21.

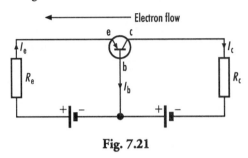

Fig. 7.21

The circuit in Fig. 7.22 is part of the circuit from Fig. 7.21 showing a forward-biased base-emitter junction.

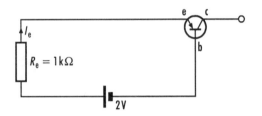

Fig. 7.22

In this circuit an emitter current flows into the emitter and out of the base. The current will be less than 2 mA because there is a potential difference across the junction of about 0.7 V if the transistor is made of silicon. The current is therefore about

$$\frac{2 - 0.7}{1000} = 0.0013\,\text{A} = 1.3\,\text{mA}$$

In Fig. 7.23 the circuit is a reverse-biased base-collector junction.

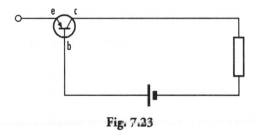

Fig. 7.23

Across this junction little or no current will flow because the width of the depletion layer increases with reverse bias as discussed in section 7.2. The minority carriers, holes in the n-type, free electrons in the p-type, will be attracted across the boundary producing a small leakage current. The circuit in Fig. 7.24 shows an $n{-}p{-}n$ transistor biased for normal operation.

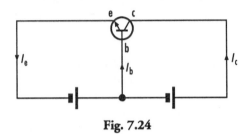

Fig. 7.24

To understand the action of an $n{-}p{-}n$ transistor the following points need to be considered.

- The base is p-type material with majority carrier holes.
- The emitter and collector are n-type material with majority carrier electrons.
- The base material is thin compared with the emitter and collector.
- Forward bias of the base-emitter junction causes electrons in the base to move toward the junction. Some recombination takes place. The collector is positively charged and thus attracts electrons from the base. A collector current I_c flows through the base-collector junction. The magnitude of this current is determined by the base-emitter junction bias.
- The base current is provided by the electrons that recombined with the holes in the base.
- From this the emitter current is the sum of the base current and the collector current, i.e.

$$I_e = I_b + I_c$$

A $p{-}n{-}p$ transistor operates in a similar manner to an $n{-}p{-}n$ transistor, but as the majority carriers flowing between emitter and collector are holes the polarities of the d.c. sources must be reversed as shown in Fig. 7.25.

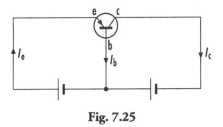

Fig. 7.25

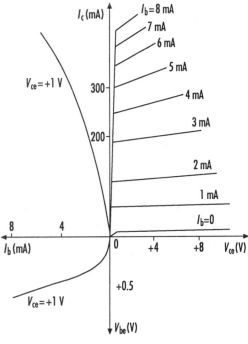

Fig. 7.27

Example

1. A $p-n-p$ transistor passes a collector current of 100 mA when the base current is 2.95 mA. Determine the value of the emitter current.

$$I_c = 100\,\text{mA} \quad I_b = 2.95\,\text{mA}$$

Using $I_e = I_b + I_c$

$$I_e = 2.95 + 100 = 102.95\,\text{mA}$$

7.10 BASIC AMPLIFIER CIRCUITS

Figure 7.26 shows the common emitter circuit, in which the common terminal is the emitter.

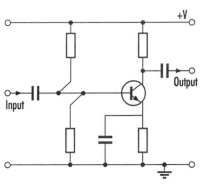

Fig. 7.26

The characteristic curves for a germanium $n-p-n$ high-gain transistor are shown in Fig. 7.27.

From the graph of I_c vs V_{ce} it can be seen that at a collector voltage above the 'knee' voltage, a comparatively large change in collector voltage produces a relatively small change in collector current. The transistor will therefore have a high output impedance.

We can also see from the graph that a small but finite collector current continues to flow, even when $I_b = 0$.

This current is the leakage current.

For a given value of V_{ce} from the graph of I_c vs I_b the current amplification is

$$\beta = \frac{I_c}{I_b} \quad \text{or} \quad h_{fe} = \frac{I_c}{I_b}$$

From the graph of I_b against V_{be} the reciprocal of the input characteristic is the input impedance.

The common emitter circuit is widely used to provide reasonably high values of input impedance, voltage output signal, power and gain.

Figure 7.28 illustrates the common base circuit, in which the common terminal is the base.

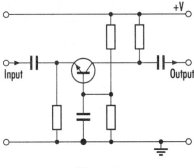

Fig. 7.28

Table 7.1

Circuit property	Common emitter	Common base	Common collector
Voltage gain	high	high	about 1
Current gain	high	about 1	high
Power gain	high	high	about 1
Input impedance	low	low	high
Output impedance	medium	high	low

A set of characteristic curves can be obtained similar to those obtained with the common emitter circuit. For a given value of V_{cb} the current amplification factor is

$$\alpha = \frac{I_c}{I_e} \quad \text{or} \quad h_{fb} = \frac{I_c}{I_e}$$

When the collector is open-circuited, $I_c = 0$, and the transistor is reduced to a base-emitter diode. The current which flows when this p-n junction is reverse biased, i.e. with the emitter negative, is the emitter leakage current. The common base circuit has a low value of input impedance but provides reasonably high values of voltage and power gain. The common collector circuit, illustrated in Fig. 7.29, is used only when the requirement is for a high input impedance and a low output impedance; the voltage gain is a little less than unity.

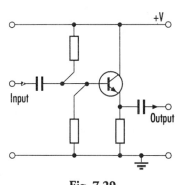

Fig. 7.29

A number of important features for the three circuits are compared in Table 7.1.

Examples

1. Calculate the current amplification factor (h_{fe}) for a common emitter circuit when the base current is 50 μA and the collector current is 2 mA at a fixed value of $V_{ce} = 6$ V.

$$I_c = 2\,\text{mA} = 2 \times 10^{-3}\,\text{A}$$

$$I_b = 50\,\mu\text{A} = 50 \times 10^{-6}\,\text{A}$$

Using $h_{fe} = I_c/I_b$

$$h_{fe} = \frac{2 \times 10^{-3}}{50 \times 10^{-6}} = 40$$

2. A transistor passes a collector current of 0.99 mA when the base current is 0.01 mA. Determine: (a) the emitter current; (b) the common base current gain (h_{fb})

$$I_c = 0.99\,\text{mA} \quad I_b = 0.01\,\text{mA}$$

(a) Using $I_e = I_b + I_c$

$$I_e = 0.01 + 0.99 = 1.00\,\text{mA}$$

(b) Using $h_{fb} = I_c/I_e$

$$h_{fb} = \frac{0.99}{1.00} = 0.99$$

3. From the graph drawn in the figure for a common emitter configuration determine, for a base-bias of 4 mA and a collector voltage of 2 V: (a) the input resistance (R_1); (b) the output resistance (R_2); (c) the current gain.

(a) From the graph I_b *vs* V_{be} at $I_b = 4\,\text{mA}$ the input resistance can be determined.

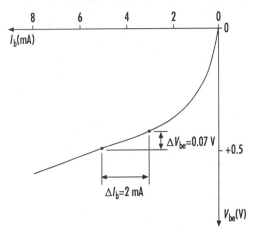

$$\Delta V_{be} = 0.07\,\text{V} \quad \Delta I_b = 2\,\text{mA}$$

Input resistance $R_1 = \Delta V_{be}/\Delta I_b$

$$R_1 = \frac{0.07}{2 \times 10^{-3}} = 35\,\Omega$$

(b) From the graph I_c *vs* V_{ce} at $I_b = 4\,\text{mA}$ the output resistance can be determined.

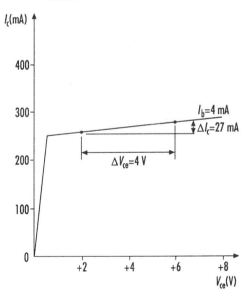

$$\Delta V_{ce} = 4\,\text{V} \quad \Delta I_c = 27\,\text{mA}$$

Output resistance $R_2 = V_{ce}/I_c$

$$R_2 = \frac{4}{27 \times 10^{-3}} = 148\,\Omega$$

(c) From the graph I_c *vs* V_{ce} at $V_{ce} = 2\,\text{V}$ the current gain can be determined.

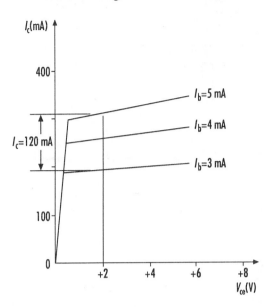

$$\Delta I_b = 5 - 3 = 2\,\text{mA} \quad \Delta I_c = 120\,\text{mA}$$

Using current gain $(h_{fe}) = I_c/I_b$

$$h_{fe} = \frac{120}{2} = 60$$

7.11 TRANSISTOR CIRCUITS

There are literally thousands and thousands of circuits that we could consider but the scope of this book will allow us to consider only a very few. Many of them will have common features that you should be able to spot time and again. This should be helpful when you begin to build such circuits.

7.11.1 Moisture detector

A circuit diagram is shown in Fig. 7.30. When the two probes detect moisture the circuit is complete. Current will flow from the positive of the battery, through the moisture and through resistor Rb, into the base of the transistor, out of the emitter of the transistor to the negative of the battery. The base current switches on the transistor collector current which lights the lamp to give a signal.

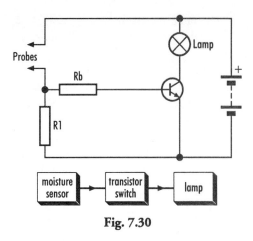

Fig. 7.30

resistor (LDR) is high and in light its resistance is low. During daylight the lamp in the circuit will not light but as soon as darkness descends the LDR resistance goes high and current flows. This current goes into the base of the first transistor, out of the emitter into the base of the second transistor and out of the emitter to the negative of the battery. The base currents switch on the transistor and, if the collector currents are large enough, the lamp lights giving as the title suggests light from darkness.

There are many applications of this circuit including detecting whether or not it is raining outside, measuring moisture content of soil, measuring body resistance and as a liquid level indicator in say a tall container on a farm or in a brewery.

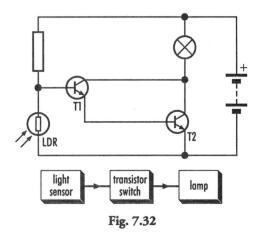

Fig. 7.32

7.11.2 Darlington pair amplifier

In the moisture-detecting circuit we used only one transistor. On some occasions this circuit is not sensitive enough to detect changes so we use two transistors called a Darlington pair. The revised circuit is shown in Fig. 7.31. This layout gives us a more sensitive circuit which means it operates for smaller changes in moisture content.

This type of circuit is often used on road-works to light up holes, etc. after darkness has fallen. Some householders use them as security lights. They are a bit expensive on batteries, but the battery can always be replaced by a power unit as discussed in Section 7.5.

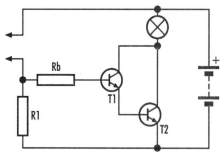

Fig. 7.31

7.11.3 Light from darkness

A circuit diagram is shown in Fig. 7.32. In darkness the resistance of a light-dependent

7.11.4 Heat-operated switch

A circuit diagram is shown in Fig. 7.33. The resistance of a thermistor decreases when it gets hot. During warm or hot conditions the thermistor resistance will drop and by comparison the circuit will have been designed so that the value of resistor R will be high. Current will flow through resistor Rb into the base of the first transistor, through the emitter into the base of transistor two, out of the emitter and to the negative of the battery. The current has been amplified, so if it is large enough it will light the signal lamp and indicate that a fire may be in progress.

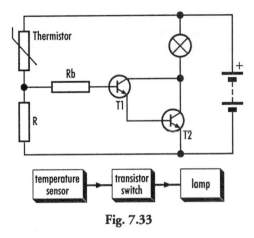

Fig. 7.33

Wherever temperature control is needed this basic circuit is available.

7.11.5 Low-temperature alarm

A circuit diagram is shown in Fig. 7.34. You should notice that in this circuit the thermistor and resistor have changed places as compared to the previous circuit. The resistance of the thermistor increases when it gets cold. When the temperature drops the thermistor resistance increases and by circuit design the resistance of resistor R is low, so once again current flows and the lamp lights. This time, however, the lamp operates because the temperature has dropped. This circuit is very useful to gardeners who need to be aware of greenhouse conditions. Other applications involve protecting equipment, etc. against cold conditions.

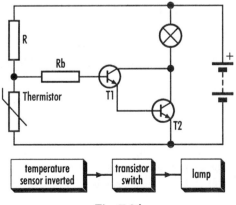

Fig. 7.34

7.11.6 Flashing signal lamps

A circuit diagram is shown in Fig. 7.35 and it is called an astable multi-vibrator. Care must be taken when wiring up the circuit to ensure that the electrolytic capacitors are connected the correct way round. The signal lamp will flash on and off because each transistor, in turn, is switched on and off. This effect is created by capacitor C1 charging and discharging through resistor R1, and capacitor C2 charging and discharging through resistor R2. The rate of flashing depends upon the time constant $T = RC$ for each set of components. Flashing signal lamps are used to indicate danger during roadworks or for pedestrian crossings and as markers at sea or on rivers and lakes.

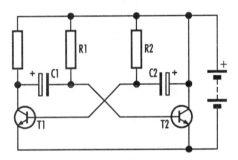

Fig. 7.35

7.11.7 Single-stage amplifier

An amplifier is a device that has an input and an output. The amplifier magnifies the input signal to produce a larger output signal. A circuit diagram for a single-stage amplifier is shown in Fig. 7.36.

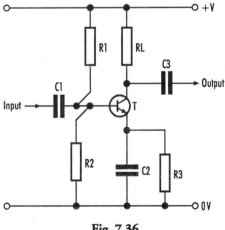

Fig. 7.36

Previous chapters have dealt with all of the components used in this amplifier. To summarize:

- the transistor T is an *n–p–n* transistor amplifier;
- resistors R1 and R2 provide a potential divider to produce the correct bias current for the base of the transistor;
- resistor RL is the load resistor across which the output voltage will be produced;
- resistor R3 is the emitter resistor which gives the correct bias for the transistor emitter;
- capacitors C1 and C3 allow an a.c. signal through but block any d.c. component of that signal – capacitors used in this way are called coupling capacitors;
- capacitor C2 is a by-pass capacitor.

7.12 INTEGRATED CIRCUITS

We have dealt with pure resistors, capacitors and inductors. These are components that do not supply energy to a circuit and are given the general name passive components. By contrast devices like diodes and transistors that control, modulate or amplify the flow of energy are called active components. An integrated circuit (IC) is a combination of interconnected active and passive devices. The advantages of having integrated circuits are:

- a reduction in costs because a great number of components can be manufactured on one small slice of silicon;
- a considerable reduction in size, volume and weight;
- that they consume less power than their equivalent passive and active components;
- that they operate faster than their equivalent passive and active components.

There are two basic types of integrated circuit:

(a) *linear or analogue*   these include amplifier-type circuits the most common of which is the operational amplifier.

(b) *digital*   these include switching circuits which have inputs and outputs that are either high or low. Digital circuits include logic gates, memories, microprocessors, etc. Digital integrated circuits divide into two types:

- *Transistor–transistor–logic (TTL)*   using bipolar transistors. Basic properties are:

  Supply voltage $5\,V \pm 0.25\,V$ d.c.
  Current needed   mA
  Switching speed   fast

- *Complementary metal oxide semiconductor (CMOS)*   using field effect transistors. Basic properties are:

  Supply voltage $3\,V$ to $15\,V$ d.c.
  Current needed μA
  Switching speed   slow

When integrated circuits became a practical possibility over the years a set of names appeared:

Small-scale integration	10 discrete circuits
Medium-scale integration	100 discrete circuits
Large-scale integration	1000 discrete circuits
Very-large-scale integration	10 000 discrete circuits
Super-large-scale integration	more than 10 000 discrete circuits

What name will be next?

7.12.1 Operational amplifiers

If you look in any trade catalogues you will find a vast number of operational amplifiers (op amps) which can be quite confusing. But we will start at the beginning. During the 1970s a 741 op amp was developed which is still in use today. This is as good a starting point as any. The 741 op amp contains the equivalent of 20 transistors, 11 resistors and a capacitor. The BS graphical symbol is shown in Fig. 7.37.

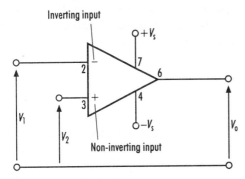

Fig. 7.37

The 741 op amp is available as a unit in both eight and 14-pin packages. The physical appearance of an 8-pin DIL op amp is shown in Fig. 7.38. The internal pin connections are shown in Fig. 7.39. Pins 1 and 5 are used as offset-null pins with pin 8 being unused. At this stage you need not concern yourself about these three pins.

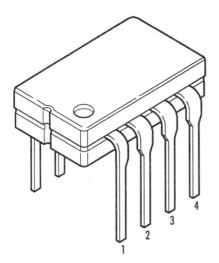

Fig. 7.38

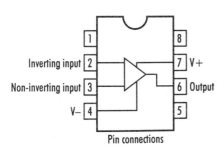

Pin connections

Fig. 7.39

The two inputs are called inverting (−) and non-inverting (+). These two signs must not be confused with the power supply. The supply voltage is usually from ±2 V to ±18 V. The basic op amp is a differential amplifier which means it amplifies the difference between the input voltages V_1 and V_2. This will give three output V_0 conditions. They are:

- when $V_2 = V_1$, the output V_0 is zero;
- when $V_2 > V_1$, the output V_0 is positive (+);
- when $V_2 < V_1$, the output V_0 is negative (−).

The open loop voltage gain is

$$\frac{V_0}{V_2 - V_1}$$

A typical characteristic graph is shown in Fig. 7.40. Over the range X–0–Y the op amp behaves in a linear fashion so the output is directly proportional to the input and there is minimum distortion of the amplifier outputs. Inputs outside the linear range cause saturation.

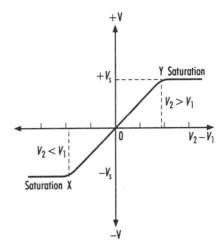

Fig. 7.40

7.12.2 555 Timer

This IC contains 25 transistors, 16 resistors and two diodes and is produced as an 8-pin DIL package. The 555 timer connections are shown in Fig. 7.41.

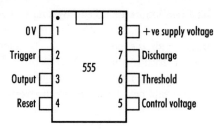

Fig. 7.41

Timers are available with a supply voltage of 2 V to 18 V. In electronic systems timing can be controlled by using either an astable circuit or a monostable circuit.

(a) Astable circuit

The circuit diagram is shown in Fig. 7.42. The square-wave output oscillations are produced automatically and depend upon the values of resistors R1 and R2 and capacitor C being used in the frequency formula. The frequency in hertz is calculated using:

$$f = \frac{1}{(R_1 + 2R_2)C}$$

where the resistance is in ohms and the capacitance is in farads. The oscillations occur because capacitor C charges up through resistors R1 and R2 and when the voltage across it reaches 67% of the supply voltage, the output goes 'low'. Capacitor C now begins to discharge and when the voltage across it falls just below 33% of the supply voltage, the output goes 'high' and capacitor C recharges. This sequence of events can be repeated continuously until the supply voltage is removed.

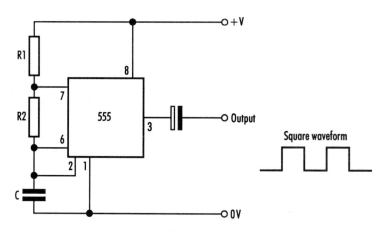

Fig. 7.42

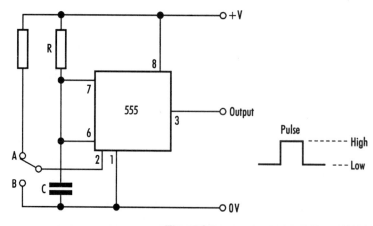

Fig. 7.43

(b) Monostable circuit

The circuit diagram is shown in Fig. 7.43. In this case we have a single square-wave output pulse produced when the switch is changed from position A to B and then back to A. This sends the output 'high' and allows the capacitor C to charge up through resistor R. When the voltage across the capacitor reaches 67% of the supply voltage, the output goes 'low' and the capacitor discharges, returning the circuit to its stable state ready for the next trigger pulse. The periodic time $T = 1.1RC$ where the resistance R is measured in ohms and the capacitance C measured in farads.

7.13 BOOLEAN ALGEBRA

Boolean algebra is a branch of symbolic logic, named after George Boole (1815–64), which is used in electronics to perform operations using a two-state condition.

7.13.1 Two-state devices

When you walk into a dark room and operate the light switch you are using a two-state device. The switch is either ON or OFF. When it is off you are using 0 V and when it is ON you are using 230 V to supply a lamp. To simplify matters we can say that we are in a '0' or '1' situation.

7.13.2 The OR operator

The addition sign + in ordinary algebra is used to represent the OR operation in Boolean algebra. We can have the following conditions:

$$0 + 0 = 0, 0 + 1 = 1, 1 + 0 = 1, 1 + 1 = 1$$

Note carefully the last of these conditions.

7.13.3 The AND operator

The multiplication sign from ordinary algebra is used to represent the AND operator in Boolean algebra. We can have the following conditions:

$$0 \times 0 = 0, 0 \times 1 = 0, 1 \times 0 = 0, 1 \times 1 = 1$$

Again note carefully the last of these conditions.

7.13.4 The NOR operator

The NOR operator is always opposite to that of the OR operator.

7.14 TYPES OF LOGIC GATES

A logic gate is the name given to a circuit which may have more than one input but one output. The input and output signals are always in a '0' or '1' state called binary signals. For convenience binary inputs to any gate will be labelled A, B, or C, etc. and the output from the gate as F. The BS graphical symbols for the various binary logic gates are shown in Fig. 7.44.

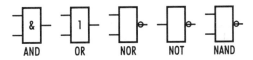

Fig. 7.44

7.15 TRUTH TABLES

With a number of inputs per gate, the input–output relationship can be expressed in the form of a truth table. A truth table lists all the possible combinations from differing input arrangements. The number of combinations can be found by using the formula:

$$\text{number of input combinations} = 2^x$$

where x = number of inputs A, B, etc.

Examples

1. Determine the number of inputs available for: (a) a 2-input gate; (b) a 3-input gate. List the inputs for each situation.

    (a) $x = 2$
    Number of input combinations = $2^x = 2^2 = 4$
    Inputs:

A	B
0	0
0	1
1	0
1	1

(b) $x = 3$

Number of input combinations $=$ $2^x = 2^3 = 8$

Inputs:

A	B	C
0	0	0
0	0	1
0	1	0
0	1	1
1	0	0
1	0	1
1	1	0
1	1	1

To complete a truth table showing the outputs available we use the rules given in Sections 7.13.

2. Given a two-input OR gate draw up a table showing the outputs.

$x = 2$. Number of input combinations $= 2^x = 2^2 = 4$
OR gate rule: use the $+$ sign from algebra. Table:

Inputs		Output
A	B	F
0	0	0
0	1	1
1	0	1
1	1	1

N.B. The output will stand on its defined 1-state if, and only if, one or more of its inputs stand at their defined 1-states.

3. Given a two-input AND gate draw up a table showing the outputs.

$x = 2$. Number of input combinations $= 2^x = 2^2 = 4$
AND gate rule: use the $\times$ sign from algebra. Table.

Inputs		Output
A	B	F
0	0	0
0	1	0
1	0	0
1	1	1

N.B. The output will stand on its defined 1-state if, and only if, ALL of the inputs stand at their defined 1-state.

4. Given a two-input NOR gate draw up a table showing the outputs.

$x = 2$. Number of input combinations $= 2^x = 2^2 = 4$
NOR gate rule: use the OR gate rule and then the NOR gate will be the opposite of the OR gate output. Table:

Inputs		Output	Output
A	B	OR gate	NOR gate
0	0	0	1
0	1	1	0
1	0	1	0
1	1	1	0

5. Complete the truth table for a three-input AND, OR, and NOR gate.

$x = 3$. Number of input combinations $= 2^x = 2^3 = 8$. Table:

Inputs			Output		
A	B	C	AND gate	OR gate	NOR gate
0	0	0	0	0	1
0	0	1	0	1	0
0	1	0	0	1	0
0	1	1	0	1	0
1	0	0	0	1	0
1	0	1	0	1	0
1	1	0	0	1	0
1	1	1	1	1	0

Answers

Exercise 7.1

1. See 7.4
2. See 7.4
3. See 7.5
4. See 7.6
5. See 7.6
6. See 7.7
7. See 7.7
8. (a) 10 mA; (b) 4 V; (c) 400 Ω; (d) 430 Ω
9. See Figs 7.4 and 7.16
10. See 7.6
11. 0.25 A or 250 mA
12. (a) 0.8333 A; (b) 7.2 Ω
13. See Fig. 7.16
14. (a) 10 V; (b) 0.02 128 A; (c) 0.02 128 A
15. Circuit (c) only

GETTING STARTED ON CONSTRUCTION

8.1 TOOLS

Before any construction work can be started we need components and tools. A basic set of tools would be:

- wire strippers
- combination pliers
- side-cutting pliers
- tapered long-nosed wire pliers
- retractable trimming knife
- soldering iron
- selection of flat blade screwdrivers
- selection of cross-point screwdrivers
- tweezers

Figure 8.1 shows in outline some of the tools listed.

As time goes by and dependent upon what path your electronics career takes you can build on this basic tool-kit. When you look in manufacturers' catalogues you will find a very wide range of tools available to cover many aspects of electronic engineering.

8.2 POWER SUPPLIES

Every electronic circuit needs a power supply so that it will work and carry out the function it was designed for. A number of alternative power supplies are available.

8.2.1 Primary cell

The purpose of a primary cell is to provide electric current to an external circuit. The single-cell voltage available is as low as 1.2 V with a capacity of 1000 mA h. Probably the most popular voltage rating is 1.5 V. A battery with a higher voltage can be made by connecting a number of cells in series. This type of battery is small and light, safe to use and easy to carry around. When any primary cell is supplying current the electrodes and the electrolyte are gradually being used up and eventually they will supply no more current. When this stage is reached a primary cell is discarded, as it cannot be regenerated.

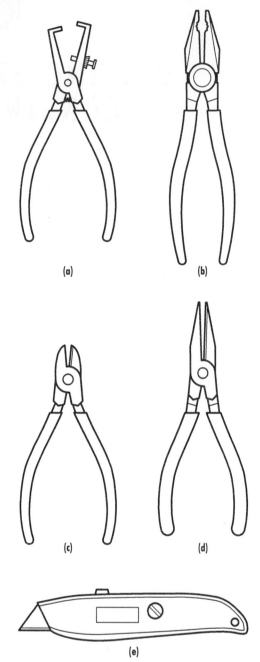

Fig. 8.1 (a) Wire strippers, (b) pliers, (c) side-cutting pliers, (d) long-nosed pliers and (e) retractable knife

A useful battery on the market today is the long-lasting lithium inorganic battery. It is used as a back-up or continuous power supply for CMOS memories, or a main power source for portable instrumentation systems. Other appli-

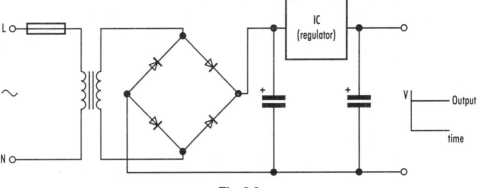

Fig. 8.2

cations include: intelligent telephones, telex machines, paging systems, seismic measurement instrumentation, data loggers, airborne navigation systems, taximeters, radio distress beacons, military coding equipment and underwater night-vision detectors. They are also used in cash registers, word processors, medical electro-cardiograph equipment, external cardiac pace-makers, hearing aids, X-ray machine control, etc.

8.2.2 Secondary cell

A secondary cell is a cell that can be recharged after use. Probably the best known secondary cell is the lead–acid cell used in the construction of motor-cycle and motor car batteries. The motor car battery is in fact six lead–acid cells. Each lead–acid cell consists of two lead plates coated in lead sulphate immersed in sulphuric acid. Another secondary cell is the alkaline cell using an alkaline electrolyte, which is usually a solution of potassium hydroxide. Each cell consists of two electrodes deposited in the electrolytic solution. To the family of alkaline cells belong: nickel–cadmium, nickel–iron, silver–zinc, silver–cadmium, and nickel–zinc. Each of the mentioned materials forms one of the electrodes. For an equivalent type of cell all of the above produce a lower e.m.f. than the lead–acid cell.

In general terms alkaline cells are more expensive than lead–acid cells for a similar capacity but they are lighter in weight yet more robust in construction than the lead–acid type of cell.

8.2.3 230 V 50 Hz mains supply

The supply voltage of the a.c. mains has recently changed to the quoted value above. In electronic circuits we usually need a d.c. voltage, so the a.c. voltage is changed to a d.c. voltage by using a rectifier circuit. Figure 8.2 shows a stabilized d.c. power supply. The point of using a stabilized d.c. power supply is that if the load current changes then the output d.c. voltage remains constant.

Figure 8.3 shows a 40 W digital d.c. power supply unit that is fully regulated and is suitable for use on a broad range of production testing, circuit design, breadboarding and educational laboratory applications. It is rated at 0–40 V at 0–1 A. It has an LED overload indicator and short-circuit protection with a reset button. It is also provided with dual d.c. digital voltmeters. Its physical size is approximately 100 mm × 400 mm × 250 mm.

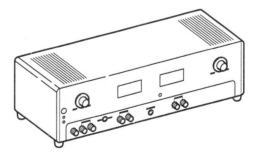

Fig. 8.3

8.3 INSTRUMENTATION

All measuring instruments are delicate and expensive and should therefore be handled with care and must be placed in a safe position so that they do not get damaged. When you are using an instrument make sure that you have read the instruction booklet provided or have been instructed by your teacher on how to use it. It is important that you use the instrument in a vertical or horizontal position as instructed by the makers and make sure that you keep the instrument away from strong magnetic fields. A magnetic field can give you a false reading which is not helpful if you are designing a circuit to a particular specification. Before using an analogue instrument adjust the pointer to zero.

A number of instruments will now be considered.

8.3.1 Ammeter

Probably the most well-known instrument of all time. It is used to measure current in amperes, milliamperes and microamperes. A good ammeter should have a low resistance so that it does not affect any circuit readings being taken. The ammeter is always connected in series with the component within a circuit.

8.3.2 Voltmeter

It is used to measure e.m.f. in volts or potential differences in volts. A good voltmeter should have a high resistance so that it does not affect any circuit readings being taken. The voltmeter is always connected across the component for the voltage to be measured.

8.3.3 Ohmmeter

It is an instrument used to determine the resistance of a circuit or a circuit component. The instrument contains a built-in source of e.m.f. and a moving coil meter movement in addition to resistors used for calibration. The principle of operation is:

1. short the test leads together;
2. adjust the built-in variable resistance until the meter reads full-scale;
3. separate the test leads and connect them to the unknown resistance;
4. read off the resistance value.

The ohmmeter scale reads in the opposite direction to other instruments, i.e. the zero reading is full-scale deflection. Figure 8.4 shows a resistance scale with a reading of 4000 Ω.

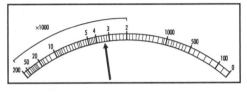

Fig. 8.4

8.3.4 Wattmeter

A wattmeter is an instrument for measuring power on both a.c. and d.c. circuits. Further details of this instrument can be found in Section 3.7.3.

8.3.5 Multimeter

This is an instrument that incorporates a voltmeter, ammeter, and ohmmeter and can therefore be set to read either a.c./d.c. voltage, a.c./d.c. current or resistance. It is available in analogue or digital form.

(a) Analogue multimeter

One popular analogue multimeter has two sets of switches as shown in Fig. 8.5. Before connecting the multimeter into the circuit it is essential that the correct setting on the selector switches is made. The first choice must be that of a.c. or d.c. In Fig. 8.5 the setting is a.c. The left-hand selector switch is now finished with. On the right-hand switch we must now select resistance, current or voltage and the most suitable numerical range. If in doubt always switch to the highest range first; when the multimeter is in circuit you can always switch down, range by range, until you find the most suitable one for your purposes. In Fig. 8.5 the setting is 10 A.

Modern analogue multimeters are fused on all ranges and have a high speed electromechanical overload cut-out for safety reasons. The sensitive cut-out, with positive latching action, is triggered by the pointer movement when overloaded. The multimeter has to be reset after overload but is quickly ready for use.

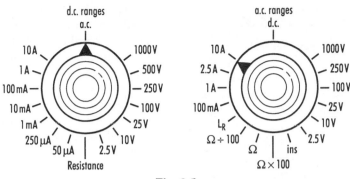

Fig. 8.5

Typical ranges are:

d.c. voltage: 100 mV, 3 V, 10 V, 30 V, 100 V, 300 V, 600 V, 1000 V
a.c. voltage: 3 V, 10 V, 30 V, 100 V, 300 V, 600 V, 1000 V
d.c. current: 50 μA, 300 μA, 1 mA, 100 mA, 1 A, 10 A.
a.c. current: 10 mA, 100 mA, 1 A, 10 A
Resistance: 0 to 2 kΩ, 0 to 200 kΩ, 0 to 20 MΩ

When current is being measured, the multimeter should be set to the highest a.c. or d.c. range and then connected in series with the component under test having first checked that the supply is switched OFF. When the multimeter is correctly connected the circuit can be switched on. The range can then be adjusted to the most appropriate scale for the measurement being taken. When voltage is being measured the same rule applies; check that the supply is OFF and that the multimeter is set to its highest range. The multimeter can now be placed in the circuit making sure that it is connected across the component. Switch on, adjust the range and then take the necessary measurement. Before placing a multimeter in circuit to measure resistance check that the supply is OFF and stays OFF. Before taking the resistance measurement it is important that the scale is calibrated otherwise a false reading will be given. Join the test leads together and adjust the pointer until it reads zero for all scales. Like an ohmmeter the zero reading is full-scale deflection. The multimeter is now ready to measure resistance. This analogue type of meter is about 150 mm × 180 mm × 80 mm deep and weighs about 2.2 kg.

(b) Digital multimeter

A much smaller and lighter multimeter is the digital hand-held type. The dimensions are approximately 135 mm × 70 mm × 30 mm with a weight of 0.5 kg.

A typical general specification is: large, easy 18 mm liquid crystal display; automatic negative polarity indication; all ranges measured by single range switch operation; low battery indicator provided; power requirements provided by a 9 V alkaline battery for 200 hours; operating temperature 0 to 40°C, etc. Typical ranges are:

d.c. voltage: 200 mV, 2 V, 20 V, 200 V, 1000 V
a.c. voltage: 200 mV, 2 V, 20 V, 200 V, 750 V
d.c. current: 200 μA, 20 mA, 200 mA, 10 A
a.c. current: 200 μA, 20 mA, 200 mA, 10 A
Resistance: 200 Ω, 2 kΩ, 20 kΩ, 200 kΩ, 2 MΩ, 20 MΩ
Capacitance: 2000 pF, 20 nF, 200 nF, 2 μF, 20 μF
Frequency: 2 kHz, 20 kHz, 200 kHz

With some hand-held multimeters you can also test diodes, take logic measurements and take continuity measurements.

As with all instruments it is important that you do not attempt to take any voltage or current measurement that might exceed the maximum range of the meter. Check this before taking any measurements and make sure that the range selected is the highest one available. This avoids any overload and you can gradually work your way down the ranges until you find an appropriate one for your measurements. With this type of instrument, always remember to switch off the power to the meter when you have finished your work.

8.3.6 Cathode ray oscilloscope (CRO)

The CRO is an instrument that enables a variety of electrical signals to be observed and examined visually. A grid of horizontal and vertical lines is fitted to the display area to enable measurements to be taken. This grid, 8 by 10 squares, is given the name graticule. Figure 8.6 shows in outline a basic CRO. Applied voltages are indicated by the deflection of a spot of light on the face of the tube. The spot records instantaneous values and can therefore be used to display a.c. waveforms when moved from left to right in a series of scans by a time-base of adjustable frequency.

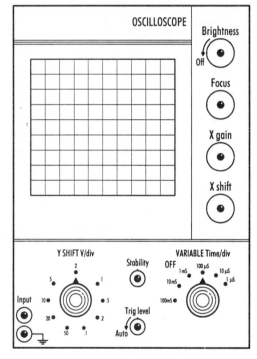

Fig. 8.6

There are a number of control knobs to be considered.

- *Brightness and focus*   These are fairly obvious in the age of computing and television receivers. Adjustments are made until a display is provided that is comfortable to your eyes.
- *Y controls*   There are two controls on the same spindle. One is calibrated in volts/division and the other moves the signal vertically up or down as the case may be.

- *X controls*   Again there are two controls, the X gain and the X shift. The X gain control allows the signal to be adjusted and expanded about the centre of the screen. The X shift control allows movement of the signal horizontally.
- *Time-base*   There are two controls on the same spindle. One is calibrated in time/division and the other ranges from off to calibrate.

The time-base is a relaxation oscillator which deflects the beam horizontally in the X direction and moves the spot across the screen from left to right. This speed of movement can be changed by using the time-base controls.

When you first come into contact with an oscilloscope it is important that you follow a simple procedure for setting it up. Before switching on:

1. set the X gain, stability control, brightness control, variable time-base control and trig level control all to fully anticlockwise;
2. set the X shift, Y shift and focus to the centre position;
3. choose the a.c. position of the a.c./d.c. switch;
4. set the time-base volts/division control to a suitable position for the signal on display;
5. set the Y control volts/division to a suitable scale for the signal under test.

The oscilloscope is now ready for use. Switch on and:

1. apply the signal to the input terminals;
2. adjust the brightness and focus;
3. adjust the X and Y shift controls until the signal is in the centre of the screen;
4. adjust the stability control until the signal is steady on the screen.

8.4 TERMINATIONS AND CONNECTIONS

A connector is a component that will connect together a number of conductors, whereas a termination is the end of a conductor prepared for connection to an accessory. A cable termination can be made by means of a terminal block, soldering socket or compression-type joint. A number of these will now be considered.

8.4.1 Terminal blocks

Glazed porcelain in 2- and 3-way terminal blocks is suitable for use in high ambient temperature situations. Inside the porcelain block is a nickel-plated brass insert and screws for placing the conductor in an appropriate position. This type of terminal block can be used in temperatures up to 500°C.

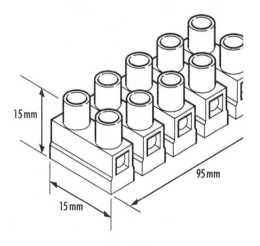

Fig. 8.7

A melamine 150 terminal block is also available. This type of block provides dimensional stability, negligible moisture absorption and high resistance to tracking. Zinc-plated steel screws and nickel-plated phosphor–bronze wire protection leaf springs are provided to clamp the conductors in the nickel-plated brass insert.

Figure 8.7 shows part of a 12-way terminal block made from self-extinguishing white nylon 6.6. The nickel-plated brass inserts have zinc-plated steel screws which screw down on to a stainless steel leaf spring. The leaf spring protects the conductors from damage on tightening. This

type of block is made in various sizes for different conductor cross-sectional areas. The dimensions in Fig. 8.7 are for a 6 A type with a maximum cable size of 1 mm^2.

By using suitable tools the conductor is prepared as shown in Fig. 8.8(a) and then placed in a one-way terminal block as shown in Fig. 8.8(b).

8.4.2 Soldered socket

This type of termination can be very small for electronic applications and very large for power applications. The cable is prepared as shown in Fig. 8.9(a) ready for insertion into the socket shown in Fig. 8.9(b). It is vitally important that the conductor, after being prepared, is completely clean and free of grease, otherwise the permanent joint that is needed will be unsatisfactory. In a like manner the socket must also be completely clean for the same reason.

One method of completing the termination is to:

1. place the flat end of the socket in a vice with the open end standing vertically upwards;
2. apply flux to the conductors,
3. nearly fill the socket with molten solder;
4. insert the conductor into the socket;
5. apply heat to the socket so that the solder remains molten;
6. remove the heat and allow the termination to cool naturally until the solder sets;
7. check that the joint is electrically and mechanically sound.

The process is now complete but during it there are a number of safety aspects to be considered. They are:

- be very careful and possibly wear protective clothing including gloves when preparing the solder bath;

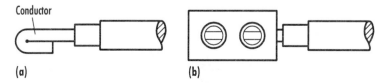

Fig. 8.8

- be careful when pouring the solder into the socket – do not allow it to splash on your body or clothing or on electric cables or on gas supplies;
- check that the conductor and socket are clean before commencement of operation;
- take care when using a naked flame, during the last operation, to avoid a risk of fire.

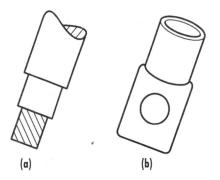

(a)   (b)

Fig. 8.9

8.4.3 Compression termination

For this operation the cable is prepared as shown in Fig. 8.10(a) and placed in the terminal as shown in Fig. 8.10(b). A crimping tool is then placed around the joint and hand pressure is applied, producing a multi-ridge pattern that results in a crimp with increased contact area and improved pull-out characteristics. A number of different types of crimping tool are available on the market, all of which can be found in manufacturers' catalogues.

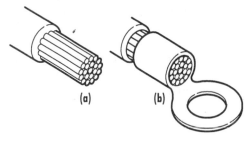

(a)   (b)

Fig. 8.10

There are many different types of terminals and connectors available. A selection are shown in Fig. 8.11.

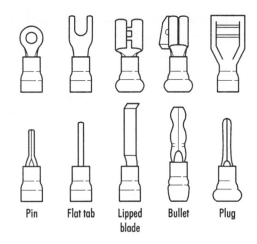

Pin   Flat tab   Lipped   Bullet   Plug
                      blade

Fig. 8.11

8.4.4 13A plug termination

In this case the conductors are formed into a loop with round-nosed pliers and placed as shown in Fig. 8.12(a). This is repeated three times for the live, neutral and earth terminal positioned as shown in Fig. 8.12(b). Care must be taken when preparing the conductors so that the direction of the loop is in the same direction as that taken by the tightening nut. If required, the loop can be made stronger by tinning it with solder but some flexibility is lost. With some 13A plug tops the connection is made by inserting the conductor into a brass screw-type single terminal block.

8.4.5 Scart connectors

This type of connector is very popular for the interconnection of television receivers, video equipment, cameras and recorders. The connector is shown in outline in Fig. 8.13. All inputs and outputs are via a 21-pin connector, the socket being PCB-mounted. The cable mounting non-reversible plug is usually supplied as a fully loaded solder-type complete with cable strain relief.

8.4.6 Other connectors

There are many other types of connectors available but they are beyond the scope of this book. When you need such a connector, go to a catalogue and make your choice. A few types are described below.

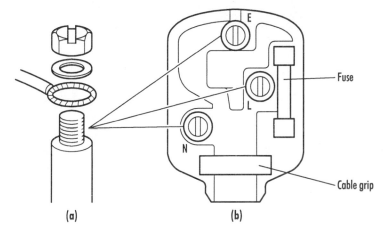

(a)   (b)

Fig. 8.12

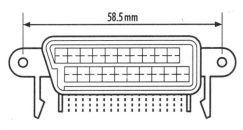

Fig. 8.13

Smart card connectors

Applications are for security access control, tele-communications, payment control in banks, machine process control, etc.

Memory card connectors

Features often include polarization which prevents incorrect card insertion and mate first, break last power, earth pins.

Direct PCB edge connectors

As the name suggests they are used on PCBs.

Insulation displacement connector (IDC)

IDCs are used to connect ribbon cable. Many suppliers offer specialized services so that virtually any length of cable can be prepared in a short space of time. As usual a plug and socket are required. In the connector there are a number of small blades that cut through the ribbon cable insulation and allow the conductors to make contact with the terminals.

DIL connectors and sockets

DIL connectors are used to make permanent connections to a PCB and the socket connects the IC to the PCB. There are a variety of types available.

Zero insertion force (ZIF) sockets

They are used for connecting IC chips to PCB. If you look at a ZIF socket you will see a small lever. The lever, when moved, gives the zero force and is usually in the locking position when down.

8.5 CONSTRUCTION METHODS

A circuit diagram is a diagram which shows by means of symbols the components and their interconnections used in the operation of a circuit. We have now reached the stage where we can take the circuit diagram and convert it to a practical wiring diagram. The wiring diagram shows the connections between components and indicates their physical layout. A number of choices are available from temporary fixing to permanent fixing. Each type will be dealt with.

8.5.1 Twisted leads

A very simple and temporary method of connecting components together is by joining leads together in a twisting motion. The result looks very messy but it will give an indication as to whether or not the circuit will work. The advantage is that it is quick. The disadvantages are that the leads become untwisted, there is poor

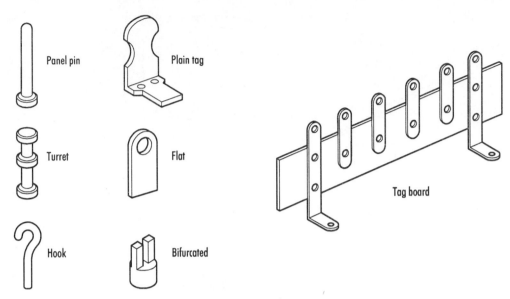

Fig. 8.14

electrical contact and that unless you are very careful short circuits occur. This method is only recommended for the very simplest of circuits.

8.5.2 Board and pin

A piece of insulated board is used and on it are mounted terminals. The size of the board can vary according to the size of the circuit it is being used for. A typical size is 150 mm × 200 mm. The types of terminal available and a tag board are shown in Fig. 8.14. The terminals can be sited on the board at any desired position and are then used to make soldered joints at con-

ductor junctions or to secure components. Each terminal can have the conducting wire wrapped around it in slightly different ways as shown in Fig. 8.15.

With the panel pin terminal the wire is wrapped half way round the circumference of the pin and soldered. The turret terminals are usually used to support components and have connecting wires. The components are placed in the upper position and the connecting wire in the lower position. With the hook terminal the conducting wire is simply hooked on to the terminal. With the plain tag terminal the wire is wrapped around the outside of the terminal

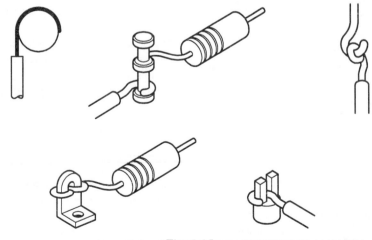

Fig. 8.15

and soldered in position. The tag board and the flat terminal have the conducting wire through the hole. Pliers are then used to bend the wire round the tag until it goes back to its starting point. One bifurcated wire-wrapped terminal is shown but other positions are also available when needed for differing circumstances.

8.5.3 Matrix board

A matrix board is about 150 mm × 120 mm × 2 mm thick and is made from an insulated material. On it are holes into which go press-fit terminal pins which can be single-sided or double-sided. This is a much quicker method than the board and pin for constructing a circuit. The soldering techniques are as before.

8.5.4 Commercial kits

There are a number of kits available but at this level of operation one of the most useful is called a Locktronics kit. It consists of a fairly substantial base-board on which are mounted metal sockets. Part of a Locktronics board is shown in Fig. 8.16(a) together with the underside of a connecting block showing a transistor in Fig. 8.16(b). It is quick and easy to use and gives great flexibility. You just have to be careful when pressing into position the components.

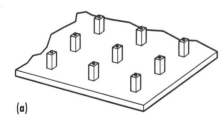

(a)

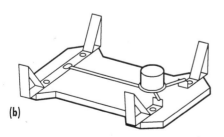

(b)

Fig. 8.16

8.5.5 Breadboard

A breadboard 150 mm × 50 mm × 10 mm thick is made from a thermoplastic material with a series of holes in it. Under each hole is a nickel–silver contact which grips a component lead or conducting wire when it is pressed into it. One such board has 47 horizontal rows of five interconnected sockets set each side of a central channel, plus two rows, one near each edge, which are used as supply rails. Part of such a board is shown in Fig. 8.17.

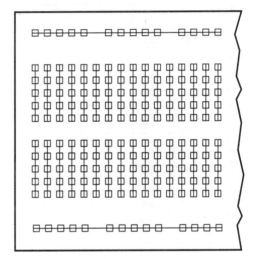

Fig. 8.17

This type of board is supplied with a bracket for the mounting of switches, etc. There is also available an extender board which means that if you have a large wiring diagram then you are able to use more than one board. To stop the unnecessary movement of the breadboard a mounting board is available that will allow three breadboards to be fixed together. You will be using breadboards during the investigations in the book.

8.5.6 Stripboard

A basic stripboard consists of a row of parallel copper strips glued to one side of an insulated board as shown in Fig. 8.18. Each strip has a number of holes in it. Boards are available in a number of sizes, one being 300 mm × 100 mm × 2 mm thick having 24 holes per strip. The components are mounted on the plain side of the board with their leads projecting through the copper strip. The leads are soldered to the copper strip and any surplus is trimmed off. It is very important to remember that the copper strip is continuous and when a connection is not needed the strip is broken. The break can be made by carefully drilling a hole with a diameter slightly larger than that of the strip width or by using a commercial cutting tool. You will be using stripboard during the investigations later in the book.

involved. The copper that you do not want removed is now protected by using an etch-resist pen.

The unprotected copper is then removed using a chemical process which leaves behind your connecting links. Before connecting the components to the PCB it is important that the board is washed and is chemical free otherwise problems may arise later. The components can be carefully soldered into position. Part of a PCB is shown in Fig. 8.19.

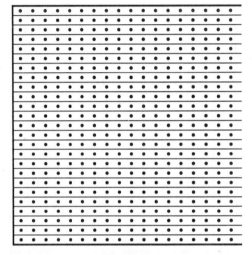

Fig. 8.18

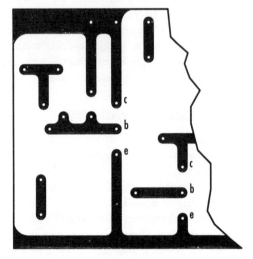

Fig. 8.19

8.5.7 Printed circuit board (PCB)

Printed circuit boards can be purchased for a particular circuit or you can make your own. A PCB is a board of insulating material covered on one side by a layer of copper. With a PCB you do not use conducting wire to connect your components together; instead you use the copper on the board. First you need to design your circuit, usually using tracing paper. Draw the components that you intend to use full size and complete the circuit with connecting lines. It is very important that all of your dimensions are accurate otherwise the components will not fit on your PCB. When you are satisfied with your diagram, you turn the tracing paper over and transfer the drawing on to the copper side of the PCB. Full details of this process are best given by your teacher, with a demonstration at the same time of how to use the chemicals

Exercise 8.1 _____

1. List the tools that you will need to form the basic tool-kit for use in an electronics laboratory.

2. State the fundamental difference between a primary and a secondary cell.

3. Write down six uses of a lithium battery.

4. State the purpose of a stabilized power supply.

5. State the purpose of an ammeter, a voltmeter, an ohmmeter and a wattmeter.

6. Draw a typical ohmmeter scale stating its main difference from that of an ammeter scale.

7. Explain with the aid of a sketch what is meant by the term 'analogue multimeter'.

8. State the difference between an analogue multimeter and a digital multimeter.

9. Explain, with the aid of a sketch, the various controls used in a CRO.

10. Explain, with the aid of a sketch, how the following terminations are made: terminal block, soldered socket, compression, 13A plug.

11. Draw an outline diagram of the following connectors: scart connector, smart card connector, memory card connector, IDC connector, DIL connector, ZIF socket.

12. Explain, with the aid of a sketch, the following types of methods of construction: twisted leads, board and pin, matrix board, breadboard, stripboard and PCB.

We have dealt with a number of electronic components and power supplies and various types of board. In this next section we will begin to construct and test circuits. We will start with fairly simple exercises and then move on to more complicated investigations.

8.6 SOLDERING EXERCISE

Aim
To mount accurately a number of types of terminal and to practise soldering.

Components and equipment needed
Baseboard, two tag boards, two panel pins, two turret terminals, two hook terminals, two plain terminals, two flat terminals, two bifurcated terminals, twelve resistors of value 100, 120, 150, 180, 220, 270, 330, 390, 470, 560, 680, and 820 Ω respectively, solder and soldering iron, standard pack of tools.

Circuit diagram
Look at Fig. 8.20(a), measure the length of the resistor wire leads from end to end and determine the distance needed between each terminal that will allow enough wire to be left over for soldering.

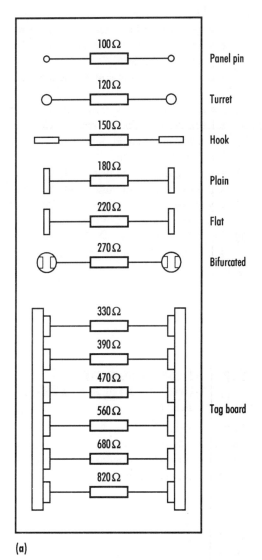

(a)

Fig. 8.20

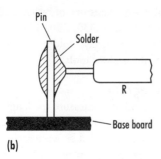

(b)

Fig. 8.20 (*cont.*)

Procedure

(a) Fix the tag board and terminals on to the baseboard as shown in diagram. Check that all terminals are clean.

(b) Place the resistor wire around the terminals in the order shown.

(c) Solder each joint as instructed by your teacher. A good joint has very low electrical resistance and should look like the one shown in Fig. 8.20(b).

(d) After a discussion with your class teacher and an inspection of your results carry out any remedial work that may be needed until you have a perfect set of soldered resistors.

(e) Keep this board in a safe place because we will be using it in future exercises.

Circuit diagram

Look at Fig. 8.21, measure the lengths of the resistor and diode leads from end to end and determine the distance between each terminal that will allow enough wire to be left over for soldering.

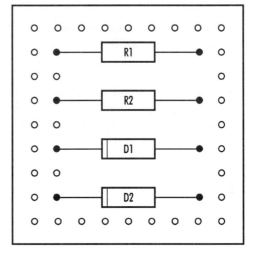

Fig. 8.21

8.7 USE OF PRESS-FIT TERMINALS IN A MATRIX BOARD

Aim

To mount accurately a number of press-fit terminals and solder components into position and then take measurements.

Components and equipment needed

Matrix board, eight press-fit terminals, two resistors of value 200 and $300\,\Omega$ respectively, two p–n diodes, digital multimeter, analogue multimeter.

Procedure

(a) Fix the press-fit terminals into the matrix board as shown in the diagram. Check that all terminals are clean.

(b) Solder the resistors and diodes into place.

(c) Adjust both multimeters for resistance measurement and adjust to zero.

(d) Take the digital meter and measure the resistance of each resistor in turn and record your results in the table.

(e) Take the analogue meter and measure the resistance of each resistor in turn and record your results in the table.

(f) Take the digital meter and measure the resistance of each diode in turn. Reverse the leads to the diode and take the measurements again.

Record your results in the table.

Table of results

	Resistor 1	Resistor 2
Nominal value		
Digital meter		
Analogue meter		

	Diode 1	Diode 2
Make/No.		
High resistance		
Low resistance		

After a discussion with your class teacher and an inspection of your results answer the following questions:

1. State what is meant by the nominal value of a resistor.

2. State the nominal value of each resistor.

3. What is the tolerance of each of the resistors?

4. Calculate for both resistors the upper and lower value of resistance.

5. Explain why there is a difference for each of the resistance measurements for each of the resistors.

6. Explain why for each diode there is a high resistance value and a low resistance value.

7. State the basic function of a diode.

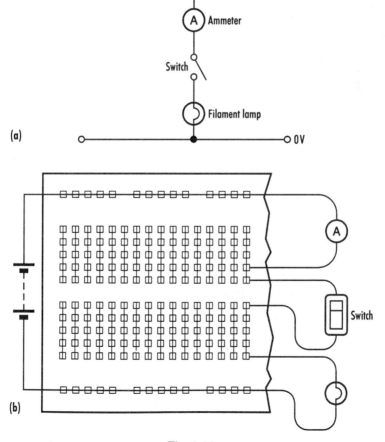

Fig. 8.22

8. State practical examples of the use of a diode.

9. A p–n junction is connected to a 15 V d.c. supply via a series resistor of 510 Ω. The current flowing is 20 mA. Calculate: (a) potential difference across the resistor; (b) potential difference across the diode; (c) power dissipated in the diode; (d) forward-bias resistance.

8.8 MEASURING CURRENT IN A FILAMENT LAMP

Aim
To use a breadboard and take measurements of electric current.

Components and equipment
Breadboard, filament lamp, ammeter and a switch.

Circuit diagrams
Look at the circuit diagram in Fig. 8.22(a) on page 149 and connect it up on the breadboard provided so that it looks like Fig. 8.22(b). Long leads will need to be used for the ammeter because it is too big to fit on the breadboard.

Procedure
(a) After wiring the circuit diagram on to the breadboard get the circuit checked before connecting the supply.
(b) Switch on and note the ammeter reading. Switch off.
(c) Disconnect the ammeter and reconnect it the other side of the switch, i.e. between the switch and the lamp.
(d) Switch on and note the ammeter reading. Switch off.
(e) Disconnect the ammeter and reconnect it on the other side of the lamp, i.e. between the lamp and negative rail.
(f) Switch on and note the ammeter reading. Switch off.

Record your results in the table.

Table of results

Ammeter position	Ammeter reading
1	
2	
3	

After a discussion with your class teacher and an inspection of your results answer the following questions.

1. Are the components connected in series or parallel?

2. Write down the current measured in each part of the circuit.

3. Comment on the pattern of the results obtained.

4. Complete the circuit diagrams (Fig. 8.23) to show the value of current at each position.

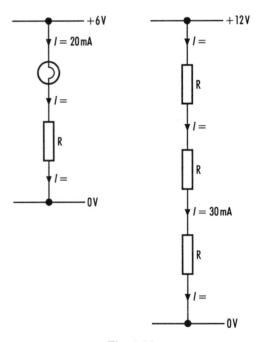

Fig. 8.23

Cell 1   Cell 2

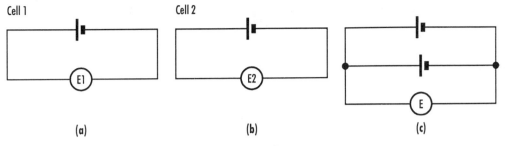

Fig. 8.24

(a)

Circuit	Cell e.m.f.	Lamp brightness
(b)		
(c)		

Fig. 8.25

8.9 CELLS CONNECTED IN PARALLEL

Aim
To establish the effect of cells connected in parallel.

Components and equipment needed
Two cells and a voltmeter.

Circuit diagrams
Fig. 8.24.

Procedure
(a) Measure the e.m.f. of cell 1 with a voltmeter as shown in Fig. 8.24(a). Note the meter reading.

(b) Measure the e.m.f. of cell 2 with a voltmeter as shown in Fig. 8.24(b). Note the meter reading.

(c) Now connect the cells in parallel and measure the total e.m.f. as shown in Fig. 8.24(c). Note the meter reading.

Table of results

	Voltmeter readings
Cell 1	
Cell 2	
Cells in parallel	

After a discussion with your class teacher and an inspection of your results answer the following questions.

1. When the cells are connected in parallel what did you notice about the results.

2. Complete the statement:
   When similar cells are connected in parallel the e.m.f. of the battery is the .... of an individual cell.

3. A filament lamp is lit by one 2 V cell to its maximum brightness as shown in Fig. 8.25(a). Now look at the table and complete each of the boxes shown. Assume identical cells are used and the lamp does not burn out.

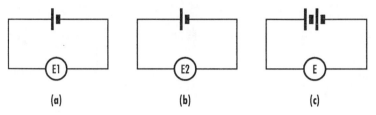

Fig. 8.26

8.10 CELLS CONNECTED IN SERIES

Aim
To establish the effect of cells connected in series.

Components and equipment needed
Two cells and a voltmeter.

Circuit diagram
Fig. 8.26.

Procedure
(a) Measure the e.m.f. of cell 1 with a voltmeter as shown in Fig. 8.26(a). Note the meter reading.
(b) Measure the e.m.f. of cell 2 with a voltmeter as shown in Fig. 8.26(b). Note the meter reading.
(c) Now connect the cells in series and measure the total e.m.f. as shown in Fig. 8.26(c). Note the meter reading.

Table of results

	Voltmeter readings
Cell 1	
Cell 2	
Cells in series	

After a discussion with your class teacher and an inspection of your results answer the following questions.

1. When the cells are connected in series what did you notice about the result.

2. Complete the statement:
When similar cells are connected in series the e.m.f. of the battery is the . . . . of the individual cells.

3. A filament lamp is lit by one 2 V cell to its maximum brightness as shown in Fig. 8.27(a). Now look at the table and complete each of the boxes shown. Assume identical cells are used and the lamp does not burn out.

(a)

	Circuit	Cell e.m.f.	Lamp brightness
(b)			
(c)			

Fig. 8.27

8.11 CONDUCTORS AND INSULATORS

Aim
To identify a material as a conductor or an insulator.

Components and equipment needed
Cell, filament lamp, single-pole one-way switch, materials as listed in the table.

Circuit diagram
Fig. 8.28 (on facing page).

Procedure
(a) Connect up the components as shown in the circuit diagram of Fig. 8.28.
(b) Take each material in turn and place it across the terminals marked 1 and 2.

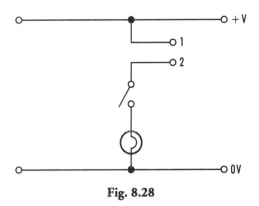

Fig. 8.28

(c) Close the switch and note the effect on the filament lamp.
(d) Record your results in the table.

Table of results

Material	Effect on lamp	Conductor/ insulator
Aluminium		
Brass		
Copper		
Lead		
Paper		
PVC		
Rubber		
Steel		

After a discussion with your class teacher and an inspection of your results answer the following questions.

1. Give an overall name to the materials that light the lamp.

2. Give an overall name to the materials that do not light the lamp.

3. What happens to the lamp if an insulating material is placed across the terminals 1 and 2?

4. What happens to the lamp if a conducting material is placed across the terminals 1 and 2?

5. Name three conducting materials and state a practical application for each one.

6. Name three insulating materials and state a practical application for each one.

7. Is the resistance in a conductor high or low? Give a reason for your answer.

8. Is the resistance in an insulator high or low? Give a reason for your answer.

9. State which of the following materials are conductors and which are insulators: silver, mercury, gold, carbon, tin, wool, platinum, tungsten, polythene, polystyrene, nylon, bakelite, polypropylene, stainless steel.

8.12 LINEAR AND NON-LINEAR COMPONENTS

Aim
To investigate the relationship between potential difference and current for; (a) a single resistor; (b) a lamp.

Components and equipment needed
Resistor, lamp, voltmeter, ammeter, single-pole one-way switch, d.c. variable supply.

Circuit diagram

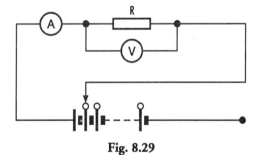

Fig. 8.29

Procedure
(a) Connect up the given components as shown in Fig. 8.29.
(b) Set the d.c. supply to its lowest value.
(c) Switch on and note the readings on the ammeter and voltmeter.
(d) Repeat stages (b) and (c) for different settings of the d.c. supply.
(e) Replace the resistor with the lamp and repeat stages (b) to (d).

Tables of results
Resistor

Voltmeter reading
Ammeter reading

Lamp

Voltmeter reading
Ammeter reading

Plot a graph for each set of results of potential difference (vertically) against current.

After a discussion with your class teacher and an inspection of your results and graphs answer the following questions.

1. State which component gives a graph with a straight line. Is this component linear or non-linear?

2. State which component gives a graph that is not a straight line. Is this component linear or non-linear?

3. State which graph obeys Ohm's law.

4. Comment upon any difficulties you encountered during this investigation.

5. The potential difference applied to a component is 100 V and the current flowing is 2 A. Calculate the resistance of the component.

6. The current in a circuit is 4 A and the resistance is 0.5 Ω. Determine the potential difference.

7. A 230 V d.c. supply is applied across a 2 Ω resistor. Calculate the current flowing through the circuit.

8. Calculate the resistance of a coil with a potential difference across it of 12 V and a current flow of 750 mA.

9. A potential difference of 230 V d.c. is applied across the terminals of a 2 kΩ resistor. Determine the current flowing through the resistor.

10. Calculate the potential difference across a resistor if its resistance is 1.5 MΩ and the current flowing through it is 20 μA.

11. Complete the table:

Potential difference (V)	230		100	12	
Current (A)	5	0.5		0.04	4
Resistance (Ω)		100	62		15

8.13 RESISTORS CONNECTED IN SERIES

Aim
To show the relationship between single resistors and a number of resistors connected in series.

Components and equipment needed
Two resistors and an ohmmeter.

Circuit diagrams
Fig. 8.30.

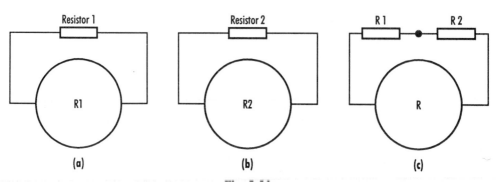

(a)   (b)   (c)

Fig. 8.30

Procedure

(a) Measure the resistance of resistor 1 using the ohmmeter as shown in Fig. 8.30(a). Note the meter reading.

(b) Measure the resistance of resistor 2 using the ohmmeter as shown in Fig. 8.30(b). Note the meter reading.

(c) Connect the two resistors in series and measure the total resistance using the ohmmeter as shown in Fig. 8.30(c). Note the meter reading.

Table of results

Resistor	R1	R2	R
Resistance (Ω)			

After a discussion with your class teacher and an inspection of your results answer the following questions.

1. Compare the value of the total resistance with the values of the individual resistors.

2. Complete the sentence:
   To find the total resistance of resistors connected in series you must ___ all the individual resistances.

3. Calculate the total resistance of a $10\,\Omega$ resistor connected in series with a $24\,\Omega$ resistor.

4. Determine the equivalent resistance of a circuit made up of three resistors of value 20, 40 and $60\,\Omega$ respectively connected in series.

5. Calculate the total resistance of a circuit made up of four resistors of value 15, 18, 24 and $30\,\Omega$ connected in series.

6. Two resistors connected in series give a total resistance of $900\,\Omega$. If one resistor is of value $470\,\Omega$ determine the value of the other.

7. Calculate the total resistance when four resistors each of $160\,\Omega$ are connected in series.

8. Resistors of value $330\,\Omega$ and $750\,\Omega$ are connected in series. Calculate the value of a third resistor which will give a total resistance of $1.2\,k\Omega$.

9. How many $0.68\,\Omega$ resistors must be connected in series to make a total resistance of $4.08\,\Omega$?

10. A filament lamp has a resistance of $36\,\Omega$. What is the resistance of 12 such lamps connected in series? How many lamps would be needed to make a total resistance of $756\,\Omega$?

11. A motor has four field coils, connected in series, each of resistance $24\,\Omega$. Calculate the total resistance of the motor. An additional series resistor is now connected giving a new total resistance of $107\,\Omega$. What would be the value of this additional resistor?

12. Two resistors of value $910\,\Omega$ and $1.2\,k\Omega$ are connected in series. Calculate the total resistance.

13. Three resistors of value $820\,\Omega$, $1.3\,k\Omega$ and $1.5\,M\Omega$ are connected in series. Determine the total resistance.

8.14 RESISTORS CONNECTED IN PARALLEL

Aim

To show the relationship between single resistors and a number of resistors connected in parallel.

Components and equipment

Two resistors and an ohmmeter.

Circuit diagrams

Fig. 8.31.

Procedure

(a) Measure the resistance of R1 using the ohmmeter as shown in Fig. 8.31(a). Note the meter reading.

(b) Measure the resistance of R2 using the ohmmeter as shown in Fig. 8.31(b). Note the meter reading.

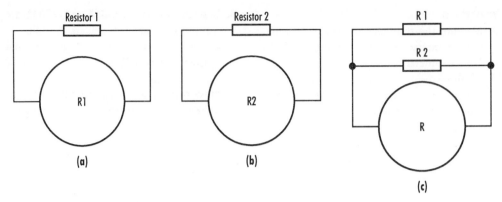

Fig. 8.31

(c) Connect the two resistors in parallel and measure the total resistance using the ohm-meter as shown in Fig. 8.31(c). Note the meter reading.

Table of results

Resistor	R1	R2	R
Resistance (Ω)			

After a discussion with your class teacher and an inspection of your results answer the following questions.

1. Compare the value of the total resistance with the values of the individual resistors.

2. Compare $1/R$ with $1/R_1 + 1/R_2$.

3. Calculate the total resistance of a $2\,\Omega$ resistor connected in parallel with a $6\,\Omega$ resistor.

4. Three resistors of value 10, 20 and $100\,\Omega$ are connected in parallel. Determine the total resistance.

5. Two resistors of value 8 and $16\,\Omega$ are connected in parallel. Calculate the total circuit resistance.

6. Determine the equivalent resistance of a circuit made up of three resistors of value 30, 60 and $90\,\Omega$ respectively connected in parallel.

7. Calculate the total resistance of a circuit made up of four resistors of value 10, 15, 30 and $120\,\Omega$ connected in parallel.

8. Two resistors connected in parallel give a total resistance of $2\,\Omega$. If one resistor is of value $6\,\Omega$ determine the value of the other.

9. Calculate the total resistance when four resistors each of value $30\,\Omega$ are connected in parallel.

10. Three resistors connected in parallel give a total resistance of $4\,\Omega$. Determine the value of the third resistor if the others have values of 15 and $12\,\Omega$.

11. Four resistors of the same value are connected in parallel. If the total resistance is $6\,\Omega$ what is the value of each resistor?

12. Two resistors of value $820\,\Omega$ and $1\,\mathrm{k}\Omega$ are connected in parallel. Calculate the total resistance.

13. Three resistors of value $910\,\Omega$, $1.2\,\mathrm{k}\Omega$ and $1\,\mathrm{M}\Omega$ are connected in parallel. Determine the total resistance.

14. Three resistors of $1.5\,\Omega$, $2\,\Omega$ and $3\,\Omega$ are connected in parallel to a $6\,\mathrm{V}$ supply. Draw a circuit diagram and calculate: (a) total resistance; (b) total current; (c) current flowing through each resistor.

8.15 EFFECTS OF AN ELECTRIC CURRENT 1

Aim
To observe the heating effect of an electric current.

Components and equipment
Battery, ammeter, fuse, single-pole one-way switch and resistance wire.

Circuit diagram

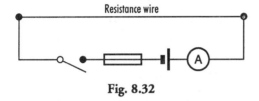

Fig. 8.32

Procedure
(a) Connect up the circuit as shown in Fig. 8.32.
(b) Close the switch for two minutes.
(c) Note the effect of the current on the resistance wire.
(d) Switch off.

After a discussion with your class teacher answer the following questions.

1. State what happened to the resistance wire; (a) 30 s after the switch was closed; (b) 1 min after the switch was closed; (c) 2 min after the switch was closed.

2. If the value of the current was doubled what would be the effect on the wire?

3. State the type of energy that was supplied to the wire.

4. After the switch was closed what type of energy was in use?

5. State a number of practical applications for this effect. The first one is completed for you.
   (a) Element in an electrical toaster
   (b) ....
   (c) ....
   (d) ....
   (e) ....
   (f) ....

8.16 EFFECTS OF AN ELECTRIC CURRENT 2

Aim
To observe the magnetic effect of an electric current.

Components and equipment
Battery, ammeter, fuse, single-pole one-way switch, dish of iron components and a coil wrapped round a former.

Circuit diagram

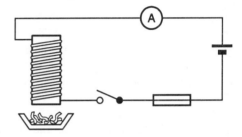

Fig. 8.33

Procedure
(a) Connect up the circuit as shown in Fig. 8.33.
(b) Switch on. Note the effect on the iron components.
(c) Switch off.

After a discussion with your class teacher answer the following questions.

1. State what happened to the components in the dish when the switch was closed.

2. What would be the effect on the iron components in the dish if the current was doubled?

3. State the type of energy being supplied to the coil of wire.

4. After the switch was closed what type of energy was in use?

5. State a number of practical applications for this effect. The first one is completed for you.
   (a) A solenoid in a motor car
   (b) ....
   (c) ....
   (d) ....
   (e) ....

8.17 EFFECTS OF AN ELECTRIC CURRENT 3

Aim
To observe the chemical effect of an electric current.

Components and equipment
Battery, ammeter, fuse, single-pole one-way switch, two copper plates, beaker containing copper sulphate solution.

Circuit diagram
Fig. 8.34.

Procedure
(a) Inspect the two copper plates and note their condition.
(b) Connect up the circuit as shown in Fig. 8.34.
(c) Switch on for five minutes.
(d) Switch off. Carefully remove the plates from the solution.
(e) Inspect the two copper plates and note their condition.

Results
Initial condition of plates

....
....
.... _____

Final condition of plates

....
....
.... _____

After a discussion with your class teacher answer the following questions.

1. What was the difference in the plates before and after the current was switched on?

2. Give a name to this effect of electric current.

3. What would be the effect on the copper plates of increasing the current?

4. State a number of practical applications for this effect. The first one is completed for you.
   (a) The chemical action taking place when a battery is in use.
   (b) ....
   (c) ....
   (d) ....
   (e) ....

5. A current of 4 A is used in charging a battery for 10 h. Determine the quantity of charge used in; (a) coulombs; (b) ampere hours.

6. Determine the total quantity discharged when a battery was used for 20 min at a rate of 10 A and then for 12 min at a rate of 8 A.

Information: consider a mass (m) in grammes (g) of a substance being liberated by a current (I) in amperes (A) flowing for a time (t) seconds (s); then $m = zIt$ where z is the electrochemical equivalent of the substance measured in grammes per coulomb (g/C).

7. A current of 6 A flows through an electrolyte of copper sulphate for 5 min. Calculate the mass of copper deposited if $z = 0.0\,003\,294$ g/C.   .

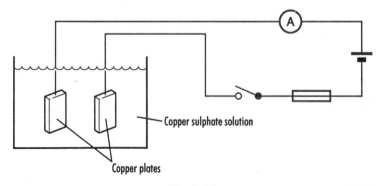

Copper sulphate solution

Copper plates

Fig. 8.34

8. Calculate the current flowing through an electrolyte if 5 g of copper are deposited on a cathode in 1.5 h.

9. Calculate the time needed to deposit 15 g of silver by a current of 8 A through a silver nitrate solution. The electrochemical equivalent for silver is 0.0 011 182 g/C.

10. A steel plate of surface area 100 cm^2 is to be nickel plated to a thickness of 0.06 mm in 50 min. Calculate the current needed if its density is 8.8 g/cm^2 and $z = 0.000\,304$ g/C.

8.18 SEMICONDUCTOR DIODE

Aim
To demonstrate how a diode acts like a switch on d.c. and like a rectifier on a.c.

Components and equipment
a.c. and d.c. supply unit, diode, single-pole one-way switch, filament lamp, CRO.

Circuit diagrams

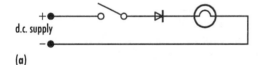

(a)

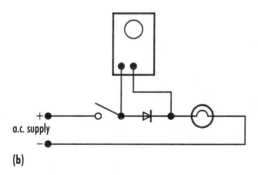

(b)

Fig. 8.35

Procedure
(a) Connect up the circuit diagram as in Fig. 8.35(a).
(b) Switch on. Note the effect on the lamp.
(c) Switch off.
(d) Reverse the connections of the diode.

(e) Switch on. Note the effect on the lamp.
(f) Switch off.
(g) Connect up the circuit diagram as in Fig. 8.35(b).
(h) Switch on and adjust the CRO so that the waveform is stationary.
(i) Draw the shape of the waveform.
(j) Switch off.

Results
Diode position 1

....

Diode position 2

....

Waveform shape

After a discussion with your class teacher answer the following questions.

1. Procedure (b). State what happened to the filament lamp.

2. Procedure (e). State what happened to the filament lamp.

3. Give an explanation of what is causing this effect in questions 1 and 2.

4. Draw a waveform for an a.c. supply.

5. Draw the waveform that was seen on the CRO from (h).

6. Explain the difference between the two waveforms.

7. Explain what is meant by the terms forward bias and reverse bias.

8. Figure 8.36 shows a circuit diagram using a diode. Will the lamp be on or off? Give a reason for your answer.

9. Figure 8.37 shows a circuit diagram using two diodes and two lamps. State which lamps will be on and which lamps will be off. Give a reason for each of your answers.

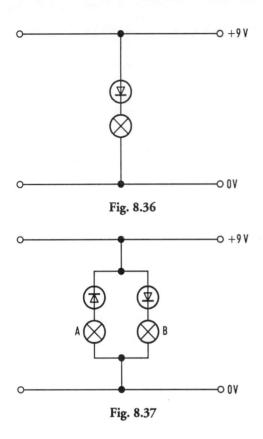

Fig. 8.36

Fig. 8.37

10. Figure 8.38 shows a circuit diagram using four diodes and five 6 V lamps. (a) Name the type of switch that is in use. (b) State the condition of the lamps when the switch is in each of the positions indicated for one minute.

8.19 A.C. FREQUENCY

Aim
To investigate the effect of frequency on resistance, capacitance and inductance.

Components and equipment
Variable-frequency oscillator, breadboard, ammeter, voltmeter, resistor, capacitor and inductor.

Circuit diagram
Fig. 8.39.

Procedure
(a) Connect up the circuit diagram as in Fig. 8.39.
(b) Switch on. Set the variable-frequency oscillator to 500 Hz.
(c) Note the frequency and the ammeter and voltmeter readings.
(d) Adjust the oscillator in 500 Hz steps and at each step note the frequency, ammeter and voltmeter readings.
(e) Continue taking readings until you reach 5000 Hz.
(f) Switch off.
(g) Replace the resistor with the capacitor and repeat steps (b) to (f).
(h) Replace the capacitor with the inductor and again repeat steps (b) to (f).

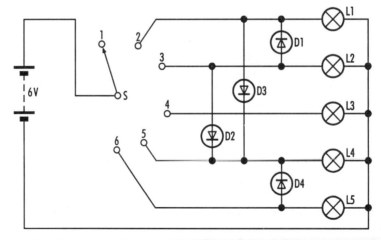

Fig. 8.38

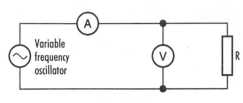

Fig. 8.39

Results
Resistor

Frequency (kHz)	0.5	1	1.5	2	2.5	3	3.5	4	4.5	5
Current										
Voltage										

Draw out a table for the capacitor and inductor results.

Calculations
- *Resistor* For each result determine the resistance using $R = V/I$.
- *Capacitor* For each result determine the capacitive reactance using $X_C = V/I$ and capacitance using $C = 1/(2\pi X_C)$.
- *Inductor* For each result determine the inductive reactance using $X_L = V/I$ and inductance using $L = X_L/(2\pi f)$.

Graphs
1. Plot graphs of voltage, vertically, against current for each components.
2. Plot graphs of resistance (R), capacitive reactance (X_C), and inductive reactance (X_L) using the vertical axis, against frequency (f).

After discussion with your class teacher answer the following questions.

1. Comment upon the shape of each graph you have drawn.

2. Write out a mathematical relationship between voltage and current for the resistor.

3. Write out a mathematical relationship between voltage and current for the capacitor.

4. Write out a mathematical relationship between voltage and current for the inductor.

5. Write out a mathematical relationship between frequency and capacitive reactance.

6. Write out a mathematical relationship between frequency and inductive reactance.

7. State what happens to resistance when frequency increases.

8. State what happens to capacitance when frequency increases.

9. State what happens to inductance when frequency increases.

10. A resistor has a current of 4 A flowing through it and a potential difference of 12 V across it. Calculate its resistance.

11. Calculate the reactance of a 200 μF capacitor when it is connected to a 230 V 50 Hz supply.

12. A capacitor has a reactance of 175 Ω when connected to a 230 V 50 Hz supply. Calculate the value of the capacitor.

13. The potential difference across a capacitor when connected to a 50 Hz supply is 230 V. Calculate the value of the capacitor if the current flowing is 12 A.

14. A coil has an inductance of 0.18 H and is being used on a 230 V 50 Hz supply. Calculate the inductive reactance of the coil.

15. A reactor has an inductive reactance of 5 Ω and is being used on a 230 V 50 Hz supply. Calculate the inductance of the reactor.

16. An inductor of 0.2 H is being used on a 100 V 50 Hz supply. Calculate the current being taken by the inductor.

17. A coil takes 5.75 A when connected to a 230 V d.c. supply and 4.6 A when connected to a 230 V 50 Hz supply. Calculate: (a) coil resistance; (b) coil reactance; (c) coil inductance.

18. State at least four applications for a capacitor and an inductor.

8.20 POTENTIAL DIVIDER

Aim

To investigate the relationship between input and output voltage using a potential divider.

Components and equipment

Variable voltage power supply unit, breadboard, three resistors, two of value 200 Ω and the third of value 300 Ω, voltmeter and a VR.

Circuit diagram

Fig. 8.40.

Procedure

(a) Connect up the circuit diagram as in Fig. 8.40(a).

(b) Switch on. Set the input voltage V_i to 2 V.

(c) Note the output voltage V_o on the voltmeter.

(d) Adjust the input voltage in steps of 2 V up to 12 V.
   At each step note the output voltage reading. Calculate V_o/V_i.

(e) Switch off.

(f) Replace resistor R2 with the 300 Ω resistor.

(g) Repeat steps (b) to (e).

(h) Connect up the VR as shown in Fig. 8.40(b).

(i) Switch on.

(j) Move the wiper of the VR from one end of the track to the other and note what happens to the voltmeter reading.

(k) Switch off.

Table of results

Circuit diagram when $R_1 = R_2 = 200\,\Omega$

Voltage V_i	2	4	6	8	10	12
Voltage V_o						
V_o/V_i						

Circuit diagram when $R_1 = 200\,\Omega$ and $R_2 = 300\,\Omega$

Voltage V_i	2	4	6	8	10	12
Voltage V_o						
V_o/V_i						

Circuit diagram using the VR
Voltmeter readings:

After a discussion with your class teacher answer the following questions.

1. What did you notice about the results for the circuit when $R_1 = R_2$?

2. How did the results differ in the second circuit as compared to the first circuit?

3. State what happened to the voltmeter readings when the VR was in use.

4. Did the first two circuits act like a VR? Give a reason for your answer.

5. Explain, with the aid of a circuit diagram, how two resistors and a VR can be used together to give greater precision in adjusting the output voltage.

Power supply unit

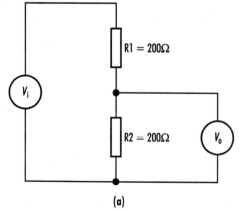

(a)

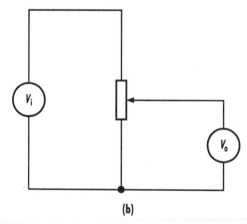

(b)

Fig. 8.40

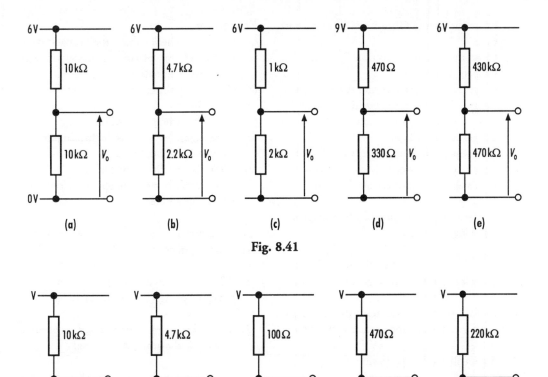

Fig. 8.41

Fig. 8.42

6. Figure 8.41 shows a number of potential dividers. In each case calculate the output voltage.

7. Figure 8.42 shows a number of potential dividers. In each case calculate the input voltage.

8.21 CHARGE AND DISCHARGE OF A CAPACITOR

Aim
To observe the effect on a signal lamp when a capacitor is charged and discharged.

Components and equipment
Power supply unit, breadboard, 2000 μF electrolytic capacitor, signal lamp, single-pole two-way switch and a timer.

Circuit diagram

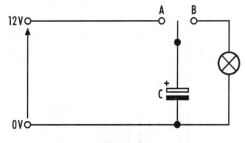

Fig. 8.43

Procedure
(a) Connect up the circuit diagram as shown in Fig. 8.43.
(b) Check that the electrolytic capacitor is connected the correct way round.

(c) Close the switch to position A for a short time.

(d) Close the switch to position B. Note what happened to the lamp.

(e) Close the switch to position A again. Open the switch to its neutral position. Wait 15 s, close the switch to position B. Note what happened to the lamp.

(f) Repeat procedure (e) in steps of 15 s for as long as your teacher tells you to do so. At each step note the effect on the lamp.

After a discussion with your class teacher answer the following questions.

1. State what happened to the signal lamp at every stage of this investigation.

2. Explain why it is important that the capacitor is connected the correct way round.

3. State a number of applications for capacitors.

8.22 VOLTAGE AND CURRENT CHARACTERISTICS IN A CHARGED AND DISCHARGED CAPACITOR

Aim
To investigate what happens to voltage and current when a capacitor is charged and then discharged.

Components and equipment
D.c. power supply unit, breadboard, 2000 μF electrolytic capacitor, 10 kΩ resistor, single-pole two-way switch, voltmeter and a centre-point ammeter.

Circuit diagram

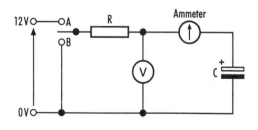

Fig. 8.44

Procedure
(a) Carefully check that the capacitor is uncharged by placing a piece of wire across its terminals.

(b) Connect up the circuit diagram as shown in Fig. 8.44.

(c) Close the switch to position A.

(d) Note carefully what is happening to the voltmeter and ammeter.

(e) Close the switch to position B.

(f) Note carefully what happens to the voltmeter and ammeter.

After a discussion with your class teacher answer the following questions.

1. When the switch is in position A, is the capacitor being charged or discharged?

2. When the switch is first closed in position A, what is the ammeter reading?

3. When the switch is first closed in position A, what is the voltmeter reading?

4. After a short time has elapsed, state what is happening to the ammeter and voltmeter readings.

5. Explain the significance of the readings taken in questions 2 and 3.

6. Explain why a centre-point ammeter is used.

7. When the switch is in position B, is the capacitor being charged or discharged?

8. When the switch is first closed in position B, what is the ammeter reading?

9. When the switch is first closed in position B, what is the voltmeter reading?

10. After a short time has elapsed, state what is happening to the ammeter and voltmeter readings.

11. Explain the significance of the readings taken in questions 9 and 10.

12. What would be the effect on the instrument readings if the value of the resistor was halved? Hint: if you are unable to answer this question repeat the investigation using the lower value resistor.

13. What would be the effect on the instrument readings if the value of the resistor was doubled? Hint: if you are unable to answer this question repeat the investigation using a higher value resistor.

14. Calculate the time constant for the circuit using the value of the capacitor and resistor provided.

8.23 INDUCTANCE AND BACK E.M.F.

Aim

To demonstrate the effect of back e.m.f. when the current through an inductor is switched off.

Components and equipment

D.c. power supply unit, breadboard, single-pole one-way switch, inductor and a neon lamp.

Circuit diagram

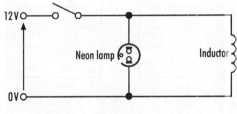

Fig. 8.45

Procedure

(a) Connect up the circuit diagram as shown in Fig. 8.45.
(b) Close the switch.
(c) Note the effect on the neon lamp.
(d) Switch off.
(e) Note the effect on the lamp.

After discussion with your class teacher answer the following questions.

1. What happened to the neon lamp when the switch was initially closed?

2. What happened to the neon lamp when the switch was opened?

3. Explain what is happening in the circuit.

4. Can this observed effect cause damage? If so, explain how it can be overcome.

8.24 DOUBLE-POLE DOUBLE-THROW SWITCH (DPDT)

Aim

To demonstrate the use of a double-pole double-throw switch.

Components and equipment

D.c. power supply unit, d.c. motor and a double-pole double-throw switch.

Circuit diagram

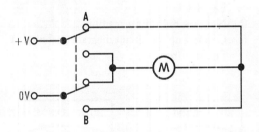

Fig. 8.46

Procedure

(a) Connect up the circuit diagram as shown in Fig. 8.46.
(b) Close the switch to position A.
(c) Note which direction the motor runs in.
(d) Switch off.
(e) Close the switch to position B.
(f) Note which direction the motor runs in.
(g) Switch off.

After a discussion with your class teacher answer the following questions.

1. When the switch was in position A which direction did the motor run?

2. When the switch was in position B which direction did the motor run?

3. What is the purpose of a double-pole double-throw switch?

4. Draw a circuit diagram using a double-pole double-throw switch for any other application that you can find.

8.25 TESTING TRANSISTORS

Aim
(a) To test a transistor to check that it is in working order; (b) to decide whether the transistor is an $n–p–n$ or a $p–n–p$ type.

Components and equipment
A number of transistors, ohmmeter, signal lamp, base resistor and a 6 V battery.

Circuit diagrams
Fig. 8.47.

Procedure
(a) Connect up the circuit diagram as shown in Fig. 8.47(a).
(b) Note the ohmmeter reading.
(c) Connect up the circuit diagram as shown in Fig. 8.47(b).

(d) Note the ohmmeter reading.
(e) Connect up the circuit diagram as shown in Fig. 8.47(c).
(f) Note the ohmmeter reading.
(g) Connect up the circuit diagram as shown in Fig. 8.47(d).
(h) Note the ohmmeter reading.

Results
Assumed $n–p–n$ transistor

Fig. 8.47(a) Emitter-base resistance = ...
Fig. 8.47(b) Base-emitter resistance = ...

Assumed $p–n–p$ transistor

Fig. 8.47(c) Base-emitter resistance = ...
Fig. 8.47(d) Emitter-base resistance = ...

Rules
$n–p–n$ transistor
Fig. 8.47(a) Resistance should be low–typically less than 1 kΩ.
Fig. 8.47(b) Resistance should be high–typically greater than 100 kΩ.

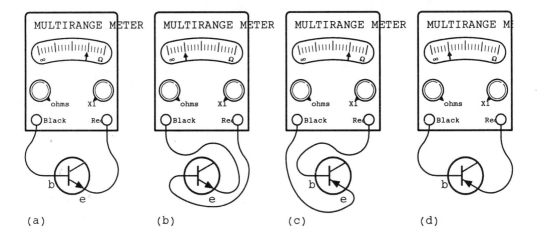

(a)   (b)   (c)   (d)

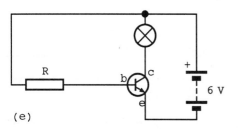

(e)

Fig. 8.47

p–n–p transistor

Fig. 8.47(c)   Resistance should be low – typically less than $1\,k\Omega$.

Fig. 8.47(d)   Resistance should be high – typically greater than $100\,k\Omega$.

Decision

You can now apply the rules to each of your transistors and decide if they are working or faulty and also decide which are n–p–n and which are p–n–p.

For n–p–n transistors a further test circuit is shown in Fig. 8.47(e). Take each of your n–p–n transistors in turn and connect up the circuit. The lamp will light if the transistor is in good condition.

8.26 SUGGESTED ANSWERS FOR INVESTIGATIONS

Section 8.7

1. The nominal value of a resistor is the value according to the colour code.
2. 200 and $300\,\Omega$ respectively.
3. Could be 5, 10 or 20%.
4. If the tolerance is 10%: R1 upper value is $220\,\Omega$ with a lower value of $180\,\Omega$; R2 upper value $330\,\Omega$ with a lower value of $270\,\Omega$.
5. The difference in resistance values is due to a wide range of values available from the nominal value. For the $200\,\Omega$ resistor its actual value for a tolerance of 20% can be between 160 and $240\,\Omega$. In addition, a further consideration to take into account is the accuracy of the multimeters.
6. When the resistance is high only a very small current will flow and the diode is reverse biased. When the resistance is low a high current will flow and the diode is forward biased.
7. The function of a diode is to be placed in a circuit so that current can only flow in one direction.
8. There are many uses for diodes. The most used is probably as a rectifier circuit in a power supply unit.

9.

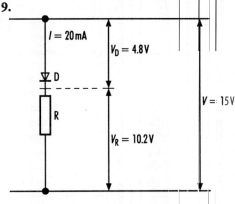

(a) $V_R = IR = 20 \times 10^{-3} \times 510 = 10.2\,V$.
(b) $V_D = 15 - 10.2 = 4.8\,V$.
(c) $P_D = V_D I = 4.8 \times 20 \times 10^{-3} = 0.096\,W = 96\,mW$.
(d) $R = V_D/I = 4.8/(20 \times 10^{-3}) = 240\,\Omega$.

Section 8.8

1. Series.
2. All current values should be the same.
3. The values are all the same confirming that the current flowing through a series circuit is the same at any point in that circuit.
4. Circuit (a) $I = 20\,mA$ at all points; Circuit (b) $I = 30\,mA$ at all points.

Section 8.9

1. All voltmeter readings are the same.
2. e.m.f.
3. Circuit 8.25(b) cell e.m.f. $= 2\,V$, lamp brightness is the same as with a single cell; circuit 8.25(c) cell e.m.f. $= 2\,V$; the brightness of each lamp is the same as that of original lamp.

Section 8.10

1. The e.m.f. of the battery is the sum of cells' e.m.f.
2. Sum.
3. Circuit 8.27(b) Cell e.m.f. $= 4\,V$; the lamp is very bright; circuit 8.27(c) cell e.m.f. $2\,V$; the brightness of each lamp is half of the original lamp.

Section 8.11

Material	Effect on lamp	Conductor/ insulator
Aluminium	On	Conductor
Brass	On	Conductor
Copper	On	Conductor
Lead	On	Conductor
Paper	Off	Insulator
PVC	Off	Insulator
Rubber	Off	Insulator
Steel	On	Conductor

1. Conductors.
2. Insulators.
3. Lamp is unlit.
4. Lamp lights.
5. Accept any three correct responses.
6. Accept any three correct responses.
7. Low – current flows easily.
8. High – current flow, if any, is difficult.
9. Conductors: silver, mercury, gold, carbon, tin, platinum, tungsten, stainless steel. Insulators: wool, polythene, polystyrene, nylon, bakelite, polypropylene.

Section 8.12

1. Resistor. Linear component.
2. Lamp. Non-linear component.
3. Resistor.
4. Difficulties could have been: connecting up circuit correctly so that instruments read in a positive manner, etc.
5. $R = V/I = 100/2 = 50\,\Omega$.
6. $V = IR = 4 \times 0.5 = 2\,V$.
7. $I = V/R = 230/2 = 115\,A$.
8. $I = 750\,mA = 0.75\,A$.
   $R = V/I = 12/0.75 = 16\,\Omega$.
9. $R = 2\,k\Omega = 2000\,\Omega$.
   $I = V/R = 230/2000 = 0.115\,A = 115\,mA$.
10. $R = 1.5\,M\Omega = 1\,500\,000\,\Omega$.
    $I = 20\,\mu A = 0.00002\,A$.
    $V = IR = 0.00002 \times 1\,500\,000 = 30\,V$.
11. $46\,\Omega$, $50\,V$, $1.6\,A$, $300\,\Omega$, $60\,V$.

Section 8.13

1. The total resistance is the sum of the individual resistances.
2. Add.

3. $34\,\Omega$.
4. $120\,\Omega$.
5. $87\,\Omega$.
6. $430\,\Omega$.
7. $640\,\Omega$.
8. $120\,\Omega$.
9. 6.
10. $432\,\Omega$, 21.
11. $96\,\Omega$, $11\,\Omega$.
12. $2110\,\Omega$ or $2.11\,k\Omega$.
13. $1\,502\,120\,\Omega$ or $1502.12\,k\Omega$ or $1.50\,212\,M\Omega$.

Section 8.14

1. There seems to be no direct connection but eventually the relationship is established as $1/R = 1/R_1 + 1/R_2$.
2. The comparison is exact.
3. $1.5\,\Omega$.
4. $6.25\,\Omega$.
5. $5.33\,\Omega$.
6. $16.36\,\Omega$.
7. $4.8\,\Omega$.
8. $3\,\Omega$.
9. $7.5\,\Omega$.
10. $10\,\Omega$.
11. $24\,\Omega$.
12. $450.55\,\Omega$.
13. $517.3\,\Omega$.
14. $0.67\,\Omega$, $9\,A$, $4\,A$, $3\,A$, $2\,A$.

Section 8.15

1. The responses will depend upon the type of wire used but if a good quality resistance wire is used then; (a) after 30 s the wire will become red and will be warm; (b) after 1 min the wire will become redder and hotter; (c) after 2 min the wire will glow and become quite hot.
2. The wire would probably burn out.
3. Electrical energy.
4. Heat energy.
5. There are many practical applications for a heating element, e.g. electric fire, cooker element, tungsten filament of a filament lamp, etc.

Section 8.16

1. The iron components would be attracted to the coil.
2. The attraction would take place quicker and more components would be attracted.
3. Electrical energy.
4. Magnetic energy.
5. Coil in a bell or buzzer or relay or CRO, etc.

Section 8.17

1. A film of copper had been taken from one plate and transferred to the other plate.
2. This is the chemical effect of an electric current.
3. With a higher current more copper would have been transferred from one plate to the other.
4. In practical terms this process is called plating and can be used to coat one material with another, e.g. copper plating, silver plating, gold plating, chromium plating. It is also the basic process used in primary and secondary cells.
5. (a) $Q = It = 4 \times 10 \times 60 \times 60 = 144\,000$ C; (b) $Q = It = 4 \times 10 = 40$ Ah.
6. Discharge $Q = I_1 t_1 + I_2 t_2 = (10 \times 20/60) + (8 \times 12/60) = 9.93$ Ah.
7. $\dot{m} = zIt = 0.0\,003\,294 \times 6 \times 5 \times 60 = 0.593$ g.
8. $I = m/zt = 5/(0.0\,003\,294 \times 1.5 \times 60 \times 60) = 2.81$ A.
9. $t = m/zI = 15/(0.0\,011\,182 \times 8) = 1676.8$ s $= 27.95$ min.
10. Thickness $= 0.06$ mm $= 0.006$ cm.
    Volume $(V) =$ surface area $\times$ thickness $= 100 \times 0.006 = 0.6$ cm^3.
    Mass $(m) =$ density $(D) \times$ volume $(V) = 8.8 \times 0.6 = 5.28$ g.
    $I = m/zt = 5.28/(0.000\,304 \times 50 \times 60) = 5.789$ A.

Section 8.18

1. Lamp will light.
2. Lamp is out.
3. The characteristic of a diode is that it will only allow current to flow in one direction.
4. 〰
5. 〰〰

6. The a.c. waveform is a full sine wave. The second waveform shows the positive half cycles when the diode is conducting.
7. The diode is forward biased if the potential difference across the diode is greater than about 0.6 V for a silicon diode, the forward current flows. The diode is reverse biased when there is no current flow.
8. The lamp will be on because the diode is forward biased.
9. Lamp A will be off because the diode is reverse biased; Lamp B will be on because the diode is forward biased.
10. (a) Six-pole rotary switch.
    (b)

Switch position	1	2	3	4	5	6
Lamp 1	off	on	on	off	off	off
Lamp 2	off	off	on	off	off	off
Lamp 3	off	off	off	on	off	off
Lamp 4	off	on	on	off	on	on
Lamp 5	off	off	off	off	off	on

Section 8.19

1. V against I: straight line graph for all components.
   Capacitor: X_c against f gives a curve; when f increases X_c increases. Inductor: X_L against f gives a straight line; when f increases X_L increases.
2. Voltage is directly proportional to current.
3. Voltage is directly proportional to current.
4. Voltage is directly proportional to current.
5. Frequency is inversely proportional to capacitive reactance.
6. Frequency is directly proportional to inductive reactance.
7. Nothing.
8. Capacitance gets smaller.
9. Inductance gets smaller.
10. $R = V/I = 12/4 = 3$ A.
11. $X_c = 1/(2\pi f C) = 1/(2\pi \times 50 \times 200 \times 10^{-6}) = 15.9\,\Omega$.
12. $C = 1/(2\pi f X_c) = 1/(2\pi \times 50 \times 175) = 0.00\,001\,819$ F $= 18.19\,\mu$F.
13. $X_c = V/I = 230/12 = 19.1\,\Omega$.
    $C = 1/(2\pi f X_c) = 1/(2\pi \times 50 \times 19.17) = 0.000\,166$ F $= 166\,\mu$F.

14. $X_L = 2\pi f L = 2\pi \times 50 \times 0.18 = 56.56\,\Omega$.
15. $L = X_L/(2\pi f) = 5/(2\pi \times 50) = 0.0159\,H$.
16. $X_L = 2\pi f L = 2\pi \times 50 \times 0.2 = 62.84\,\Omega$.
    $I = V/X_c = 100/62.84 = 1.59\,A$.
17. $R = V_{dc}/I_{dc} = 230/5.75 = 40\,\Omega$.
    $X_L = V_{ac}/I_{ac} = 230/4.6 = 50\,\Omega$.
18. Capacitor: part of a smoothing circuit, suppressors, filters, coupling, power factor correction, blocking d.c., separating a.c. from d.c., timing circuits, tuning circuits, etc.

    Inductors: part of a smoothing circuit, tuning circuits, blocking a.c., relays, filters, part of an aerial, control of current in a fluorescent tube, radio-frequency chokes, etc.

Section 8.20

1. When $R_1 = R_2$ the ratio always came out to the same value.
2. When R_2 was greater than R_1 then the ratio increased for each value of input voltage.
3. The voltmeter reading varied smoothly from $0\,V$ to a maximum.
4. In both cases the output voltage was different to the input voltage so the circuits did act as variable resistors.
5. Connect a resistor in series with the variable resistor. The potential divider will have a smaller output range but the circuit will provide greater precision in adjusting the output voltage.
6. $V_o = (V_i \times R_2)/(R_1 + R_2)$
   (a) 3 V, (b) 1.913 V, (c) 4 V, (d) 3.7125 V, (e) 4.7 V.
7. $V = V_o(R_1 + R_2)/R_2$
   (a) 18.76 V, (b) 37.64 V, (c) 14.55 V, (d) 11.76 V, (e) 29.36 V.

Section 8.21

1. With the switch in position A the capacitor is being charged. With the switch in position B the capacitor is being discharged through the signal lamp. The lamp will flash. As time goes by the capacitor will hold its charge and when discharged across the lamp, the lamp will flash on

2. The capacitor is an electrolytic which means it must be connected the correct way round otherwise it can be damaged even to the extent of exploding and sometimes damaging other circuit components.
3. There are numerous applications for capacitors mentioned in previous chapters but the main application on this occasion would be a photo-flash with a camera.

Section 8.22

1. Charged.
2. The ammeter reading will be approximately 1.2 mA if the supply was 12 V and the resistor was 10 kΩ.
   Check:
   $I = V/R = 12/10\,000 = 0.0012\,A = 1.2\,mA$.
3. 0V.
4. The ammeter reading is dropping and the voltmeter reading is rising.
5. The capacitor is starting to be charged to its maximum potential.
6. When the capacitor is being charged the current is moving in one direction and on discharge is moving in the opposite direction. If a centre-point ammeter was not used, on discharge the current flow could damage the ammeter because it would try and move backwards.
7. Discharged.
8. Approximately 1.2 mA.
9. Approximately 12 V.
10. The ammeter reading is dropping from 1.2 mA to almost zero and in the opposite direction on the ammeter to its original direction.
11. The capacitor is being discharged.
12. The time taken for the capacitor to charge and discharge would be shorter.
13. The time taken for the capacitor to charge and discharge would be longer.
14. $T = RC = 10 \times 10^3 \times 2000 \times 10^{-6} = 20\,s$.

Section 8.23

1. Nothing happened to the neon lamp.
2. The lamp began to glow.
3. An e.m.f. is generated when the switch is opened. The value of this e.m.f is signifi

cantly larger than the supply e.m.f. Since this generated e.m.f. is in the reverse direction to the applied voltage it is called a back e.m.f. The back e.m.f. occurs because of the collapse of the magnetic field when the current is switched off.

4. This reverse current flow can go through any components in its way. Such devices as transistors are easily damaged and must therefore be protected. A semiconductor diode is used to protect any number of devices. More details will be provided in later investigations about such a protective diode.

Section 8.24

1. The motor will run in a clockwise direction.
2. The motor will run in an anticlockwise direction.
3. A double-pole double-throw switch can control the direction of current flow.
4. Any number of applications are available. A vehicle could be controlled for forward and backward movement. In addition, to make the vehicle stop if it bumped into some obstruction, two limit switches could be included.

Section 8.25

The teacher can check the students' work by using the results obtained from the tests including the final test of lighting the signal lamp.

Answers

Exercise 8.1

1. See 8.1
2. See 8.2.1 and 8.2.2
3. See 8.2.1
4. See 8.2.3
5. See 8.3
6. See 8.3.3
7. See 8.3.5
8. See 8.3.5
9. See 8.3.6
10. See 8.4
11. See 8.4
12. See 8.5.1–8.5.7

9

ELECTRONIC CIRCUITS

9.1 INTRODUCTION

If you have completed our studies up to this point in a satisfactory manner you are now ready for the final set of investigations. It is suggested that all of the following investigations can be constructed using breadboard. Specific components have not been listed because it is fully accepted that different electronic laboratories have differing resources so the teacher will provide what is available to carry out each investigation. If on completion of an investigation you wish to retain the circuit you can then make it permanent by using a PCB.

You are reminded to:

1. collect all of the components you need before starting an investigation;
2. carry out tests to check that all components are in working order;
3. check which connections are which on diodes, capacitors, transistors, etc.;
4. study the circuit diagram before moving on to assembly;
5. start off by inserting the most complicated components on to the breadboard – probably an IC or transistor;
6. then insert capacitors, resistors and all sorts of diodes;
7. insert connecting wires;
8. check that all components and wires are firmly in place;
9. finally connect the supply rails of the breadboard to a power supply unit;
10. test the circuit to check that it works.

Fault finding
If the circuit does not work, switch off. Check:

- your circuit diagram;
- the wiring diagram;
- the position of all components and connecting wires.

Be logical; only change one wire or one component at a time. If all else fails, seek advice and guidance from your teacher.

9.2 LIGHT-OPERATED SWITCH USING A LIGHT-DEPENDENT RESISTOR (LDR)

Aim
To observe the effect of light on an LDR.

Components and equipment
Ohmmeter, LDR, two $n–p–n$ transistors, two resistors, electrolytic capacitor, signal lamp, battery, breadboard.

Circuit and wiring diagrams

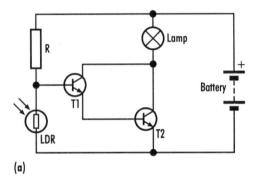

(a)

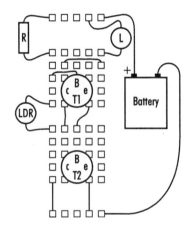

(b)

Fig. 9.1

Procedure

(a) Inspect the circuit diagram (Fig. 9.1(a)) and then connect up the wiring diagram as shown in Fig. 9.1(b).

(b) Place your hand over the LDR and note what happens to the signal lamp.

(c) With an ohmmeter measure the resistance of the LDR with and without your hand over the LDR.

(d) Replace resistor R with a resistor of twice its value and again note what happens to the signal lamp. Remove the second resistor and replace it with the original resistor.

(e) Connect the electrolytic capacitor across the LDR. Be careful to make sure that the positive lead of the capacitor is connected to the side of the LDR that is connected to the base of transistor 2. Note what happens to the signal lamp and then remove the capacitor.

(f) Remove resistor R and the LDR. Connect the LDR in the resistor R position and resistor R in the LDR position. Note what happens to the signal lamp.

After a discussion with your class teacher and an inspection of your results answer the following questions.

1. In the original circuit, procedure (b), explain what you found out about an LDR and state one practical use for such a circuit.

2. State the value of resistance when the LDR was in light and out of light.

3. Explain what happened to the signal lamp when resistance *R* was doubled in value.

4. What happens to the signal lamp when a capacitor is placed across the LDR?

5. Why is it important to take care when connecting up the electrolytic capacitor?

6. What happens to the signal lamp when the resistor and the LDR are swapped over? State one practical use for such a circuit.

7. What would be the advantage of using a variable resistor instead of a fixed resistor *R*?

8. State the purpose and the name given to the two transistors when they are connected as shown in the circuit diagram.

9. Give a detailed explanation of how the circuit works.

9.3 HEAT SENSING UNIT

Aim
To observe the effect of heat and cold on a thermistor.

Components and equipment
Negative temperature coefficient (NTC) thermistor, resistor, VR, two transistors, signal lamp, battery and breadboard.

Circuit diagram

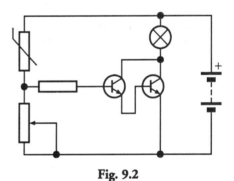

Fig. 9.2

Procedure

(a) Inspect the circuit diagram (Fig. 9.2) and then draw out a wiring diagram on a breadboard sheet.

(b) Get your teacher to check the wiring diagram and then connect up the circuit.

(c) Adjust the VR until the signal lamp is just OFF.

(d) Gently heat the thermistor and note what happens to the lamp.

(e) Allow the thermistor to return to its original temperature and then interchange the thermistor with the VR.

(f) Adjust the VR until the signal lamp is just OFF.

(g) Cool the thermistor with ice and note what happens to the signal lamp.

After a discussion with your class teacher and an inspection of your results answer the following questions.

1. State what happened to the signal lamp when the thermistor was heated.

2. Explain how the circuit works when the thermistor is heated.

3. Give at least one practical application for such a circuit.

4. State what happened to the signal lamp when the thermistor was cooled.

5. Explain how the circuit works when the thermistor is cooled.

6. Give at least one practical application for such a circuit.

7. Plot a graph of resistance (vertically) against temperature for the set of results obtained from an investigation on a thermistor.

Resistance (kΩ)										
2	4	6	8	10	12	14	16	18	20	
Temperature (°C)										
25	14	5	0	−8	−12	−18	−20	−25	−28	

(a) Use the graph to read off the resistance of the thermistor at (i) +20°C (ii) +10°C (iii) 0°C (iv) −10°C (v) −20°C. (b) Use the graph to read off the temperature of the thermistor at (i) 5 kΩ (ii) 10 kΩ (iii) 15 kΩ. (c) Determine the power used at 25°C and 8°C if the power required to raise the thermistor temperature 1°C is 0.4 mW/°C. (d) Explain why the resistance of the thermistor changes with temperature in this way.

8. Write out the name of one other type of thermistor available and explain the difference between it and the one used in your investigation.

9. A 500 mW bead thermistor has a resistance of 2.2 kΩ at a temperature of 25°C. If it is used on an applied voltage of 6 V determine

whether or not the power dissipated in the bead exceeds its rated maximum.

10. If the thermistor bead in question 9 is now used on a 35 V supply determine whether or not the power dissipated in the bead exceeds its rated maximum.

11. If a thermistor is used in a circuit where it does exceed its maximum power rating explain what is likely to happen to the thermistor.

9.4 LIQUID LEVEL DETECTOR

Aim
To build a circuit that will detect whether or not a liquid has reached its highest point in a container.

Components and equipment
Two resistors, two transistors, breadboard, signal lamp, battery, pair of probes and a container with liquid.

Circuit diagram

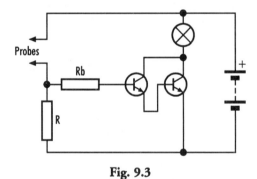

Fig. 9.3

Procedure
(a) Inspect the circuit diagram (Fig. 9.3) and then draw out a wiring diagram on a breadboard sheet.

(b) Get your teacher to check the wiring diagram and then connect up the circuit.

(c) Place the probes in a container so that they do not touch the liquid. Note the effect on the signal lamp.

(d) Add more liquid to the container until the liquid is touching the probes. Note what happens to the signal lamp.

After a discussion with your class teacher and an inspection of your results answer the following questions.

1. State what happened to the signal lamp when the liquid was not touching the probes.

2. State what happened to the signal lamp when the liquid was touching the probes.

3. Explain how the circuit works.

4. Give at least one practical application for such a circuit.

9.5 SIGNAL LAMP WARNING

Aim
To build a circuit that will act as a warning device during the hours of darkness.

Components and equipment
Two transistors, four electrolytic capacitors, three resistors, two signal lamps, breadboard single-pole one-way switch and battery.

Circuit diagram

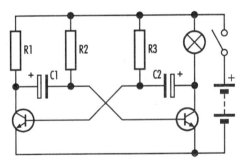

Fig. 9.4

Procedure
(a) Inspect the circuit diagram (Fig. 9.4) and then draw out a wiring diagram on a breadboard sheet.
(b) Get your teacher to check the wiring diagram and then connect up the circuit.
(c) Close the switch and note what happens to the signal lamp. After a suitable time interval open the switch.
(d) Replace capacitor C1 with a capacitor of 10% capacitor$_1$.

(e) Switch on and again note the effect on the signal lamp. Switch off and replace the capacitor with the original C1.
(f) Replace capacitor C2 with a capacitor of 10% C2.
(g) Switch on and again note the effect on the signal lamp. Switch off and replace the capacitor with the original C2.
(h) Replace resistor R1 with another signal lamp. Note what happens to the two signal lamps. Switch off.

After a discussion with you class teacher and an inspection of your results answer the following questions.

1. During procedure (c) state what happened to the signal lamp.

2. Explain how the circuit works.

3. During procedure (e) state what happened to the signal lamp.

4. Explain how the second circuit worked.

5. During procedure (h) state what happened to the two signal lamps.

6. Give a name to this special type of circuit.

7. Give at least one practical application for this circuit.

8. Explain the purpose of resistor R2.

9. Calculate the frequency of an astable multi-vibrator for the circuit diagram shown in Fig. 9.4 having the following component values:
   Capacitor $C_1 = C_2 = 0.2\,\mu\text{F}$. Resistor $R_2 = R_3 = 10\,\text{k}\Omega$. Hint: use the formula frequency $f = 1/(1.4C_1R_2)$.

9.6 SOUND ALARM

Aim
To build and investigate a circuit that will act as a sound alarm.

Components and equipment

Two transistors, four resistors, two electrolytic capacitors, LDR, loudspeaker, breadboard and a battery.

Circuit diagram

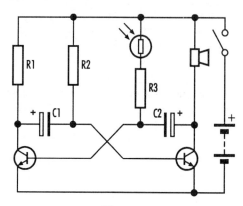

Fig. 9.5

Procedure

(a) Inspect the circuit diagram (Fig. 9.5) and then draw out a wiring diagram on a breadboard sheet.

(b) Get your teacher to check the wiring diagram and then connect up the circuit.

(c) Close the switch and note the effect on the loudspeaker. Open the switch.

(d) Cover the LDR completely so that no light can enter it. Close the switch and note the effect on the speaker. Open the switch.

After a discussion with your class teacher and an inspection of your results answer the following questions.

1. State what happened to the loudspeaker when light was entering the LDR.

2. State what happened to the loudspeaker when light did not enter the LDR.

3. Explain how the circuit works and give at least one practical application for this circuit.

4. What would be the effect on the circuit if resistor R3 was made much larger?

9.7 LAMP TIME-DELAY

Aim

To build and investigate a circuit that will allow a lamp to light some time after the switch has been closed.

Components and equipment

Two transistors, four capacitors, VR, three resistors, single-pole double-throw switch, signal lamp, stopwatch, breadboard and battery.

Circuit diagram

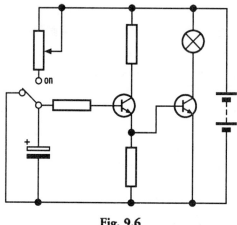

Fig. 9.6

Procedure

(a) Inspect the circuit diagram (Fig. 9.6) and then draw out a wiring diagram on a breadboard sheet.

(b) Get your teacher to check the wiring diagram and then connect up the circuit.

(c) Set the VR to its maximum resistance and switch on. Note the time the signal lamp takes to come on.

(d) Adjust the VR, in stages, to its minimum value. At each stage, note the time for the lamp to come on.

(e) Switch off. Adjust the VR to its half-way value. Switch on, and note the time for the lamp to come on. Switch off.

(f) Remove the capacitor and replace it with one of twice its value. Switch on, and note the time for the lamp to come on. Switch off.

(g) Repeat procedure (f) for two other capacitors, approximately doubling up on each occasion.

After a discussion with your class teacher and an inspection of your results answer the following questions.

1. During procedures (c) and (d) state what happened to the signal lamp. Include in your answer any measurements of time that were noted.

2. Compile a table that shows your results for procedures (e), (f) and (g). Comment upon the results.

3. Give at least one practical application for this type of circuit.

9.8 THE USE OF SEMI-CONDUCTOR DIODES IN POWER SUPPLIES

Aim
To build and investigate circuits using semiconductor diodes for power supplies.

Components and equipment
A.c. mains, transformers, fuse, four semiconductor diodes, smoothing capacitor, resistor, zener diode, breadboard, a two-pole single-throw switch and a CRO.

Part 1: single diode on a.c.

Circuit diagram
Fig. 9.7(a).

Procedure
(a) Inspect the circuit diagram (Fig. 9.7(a)) and then draw out a wiring diagram on a breadboard sheet.
(b) Get your teacher to check the wiring diagram and then connect up the circuit.
(c) Connect the CRO across the resistor, switch on, make any necessary adjustments and make a copy of the waveform that appears on the screen. Switch off.

Part 2: two diodes on a.c.

Circuit diagram
Fig. 9.7(b).

Procedure
(a) Inspect the circuit diagram (Fig. 9.7(b)) and then draw out a wiring diagram on a breadboard sheet.
(b) Get your teacher to check the wiring diagram and then connect up the circuit.
(c) Connect the CRO across the resistor, switch on, make any necessary adjustments and make a copy of the waveform that appears on the screen. Switch off.

Part 3: four diodes on a.c.

Circuit diagram
Fig. 9.7(c).

Procedure
(a) Inspect the circuit diagram (Fig. 9.7(c)) and then draw out a wiring diagram on a breadboard sheet.
(b) Get your teacher to check the wiring diagram and then connect up the circuit.
(c) Connect the CRO across the resistor, switch on, make any necesary adjustments and make a copy of the waveform that appears on the screen. Switch off.

Part 4: smoothing capacitor

Circuit diagram
Fig. 9.7(d).

Procedure
(a) Inspect the circuit diagram (Fig. 9.7(d)) and then draw out a wiring diagram on a breadboard sheet.
(b) Get your teacher to check the wiring diagram and then connect up the circuit.
(c) Connect the CRO across the resistor, switch on, make any necessary adjustments and make a copy of the waveform that appears on the screen. Switch off.

Part 5: power supply unit

Circuit diagram
Fig. 9.7(e).

Procedure
(a) Inspect the circuit diagram (Fig. 9.7(e)) and then draw out a wiring diagram on a breadboard sheet.

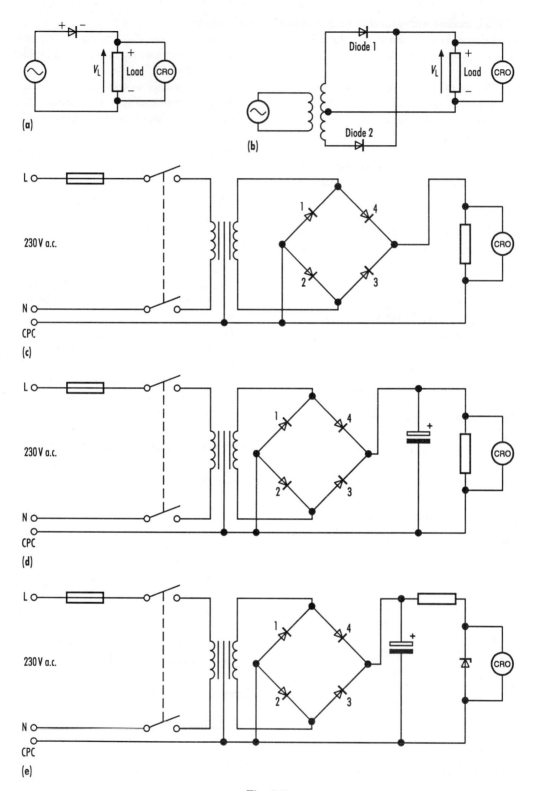

Fig. 9.7

(b) Get your teacher to check the wiring diagram and then connect up the circuit.

(c) Connect the CRO across the zener diode, switch on, make any necessary adjustments and make a copy of the waveform that appears on the screen. Switch off.

After a discussion with your class teacher and an inspection of your results answer the following questions.

1. Draw the input waveform that you would normally expect from a 230 V 50 Hz supply.

2. What is the purpose of the fuse and the single-pole double-throw switch?

3. Explain why a transformer is used in some of the circuits.

4. The input voltage to the transformer is roughly 230 V 50 Hz. State the usual approximate output voltage of the transformer.

5. A simple transformer has a primary voltage of 230 V and a secondary voltage of 20 V. If there are 1200 turns on the primary winding how many turns will there be on the secondary winding?

6. Draw each of the circuit diagrams and then add to each drawing the waveform shape that appeared on the CRO.

7. Explain for each waveform how the shape was obtained.

8. For Fig. 9.7(c) explain with the aid of sketches the conduction path through both circuits for both positive and negative half cycles.

9. Explain the need for a zener diode.

10. The zener diode in the circuit shown in Fig. 9.8 has a breakdown voltage of 4 V. (a) Calculate the current in the zener diode. (b) Determine the current in the zener diode when the supply voltage is changed to: (i) +12 V, (ii) +3 V. (c) With the supply voltage at +9 V, a resistor of resistance 100 Ω is connected between A and B. Calculate the new current in the zener diode.

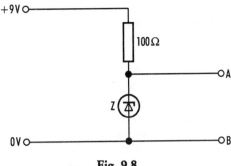

Fig. 9.8

11. The values for Fig. 9.8 are now as follows supply +11.5 V, zener diode rated at 5.2 V 500 mW, voltage at point A is 5 V. Determine; (a) maximum safe zener current; (b) potential difference across the resistor; (c) the value of the resistor. From the E12 series choose the most preferred value.

9.9 PERFORMANCE CHARACTERISTICS OF A TRANSISTOR

Aim
To plot graphs to show the characteristics of a transistor.

Components and equipment
n–p–n transistor, signal lamp, resistor, VR, two ammeters, voltmeter, breadboard and power unit.

Circuit diagram

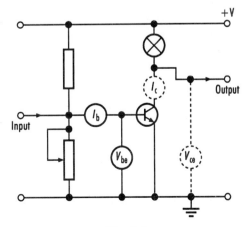

Fig. 9.9

Procedure:

(a) Inspect the circuit diagram (Fig. 9.9) and then draw out a wiring diagram on a bread-board sheet.

(b) Get your teacher to check the wiring diagram and then connect up the circuit.

(c) Switch on. Adjust the VR until a low value of base current is obtained.

(d) Increase the setting of the base current in suitable steps over its range. Measure the base current and base-emitter voltage at each interval. Switch off.

(e) Remove the voltmeter measuring base-emitter voltage.

(f) Connect an ammeter to measure the collector current and a voltmeter to measure collector-emitter voltage.

(g) Switch on. Adjust the VR until a low value of base current is obtained.

(h) Increase the setting of collector-emitter voltage in suitable steps over its range, but adjust the base current until it reads the original value for each measurement. Measure the collector-emitter voltage and collector current at each interval. Switch off.

(i) Repeat procedures (g) and (h) for another four setting of base current.

Table of results

Base current
Base-emitter voltage
Collector-emitter voltage
Collector current

Graphs

1. Plot a graph of base current (I_b) against base-emitter voltage (V_{be}).

2. Plot graphs of collector current (I_c) against collector-emitter voltage (V_{ce}).

After a discussion with your class teacher and an inspection of your results answer the following questions.

1. Comment upon the shape of the input characteristics graph of base current against base-emitter voltage.

2. Comment upon the shape of the output characteristic family of graphs of collector current against collector-emitter voltage.

3. Use the graph of I_b against V_{be} to calculate the input resistance using:

$$\text{input resistance} = \delta V_{be}/\delta I_b$$

for one value of base current.

4. Use the graph of I_c against V_{ce} to calculate output resistance using:

$$\text{output resistance} = \delta V_{ce}/\delta I_c$$

for the same value of base current chosen for the input resistance.

5. In general terms, compare the values of input and output resistance.

6. When the input current is zero, the output current should also be zero. State why this is not so in practice.

9.10 THE GAIN OF AN *N–P–N* TRANSISTOR

Aim

To measure the gain of a number of *n–p–n* transistors.

Components and equipment

Four identical *n–p–n* transistors, two resistors, VR, two ammeters, breadboard and power unit.

Circuit diagram

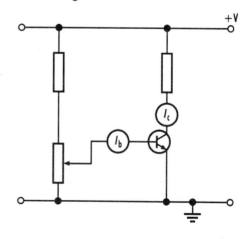

Fig. 9.10

Procedure

(a) Inspect the circuit diagram (Fig. 9.10) and then draw out a wiring diagram on a breadboard sheet.

(b) Get your teacher to check the wiring diagram and then connect up the circuit.

(c) Switch on. Adjust the VR until a low value of collector current is obtained.

(d) Increase the setting of the collector current in suitable steps over its range. Measure the collector current and base current at each interval. Switch off.

(e) Repeat procedures (c) and (d) for the other three transistors.

Table of results

Repeat this table for each of the four transistors.
Transistor

Collector current	
Base current	
Current gain $= I_c/I_b$	

After a discussion with your class teacher and an inspection of your results answer the following questions.

1. Comment upon the four sets of results you have obtained.

2. Go to a manufacturer's data sheet and write down the stated current gain for your transistors.

3. All four n–p–n transistors have the same code number but do they have the same gain? Give a reason for your answer.

4. Calculate the current amplification factor for a common emitter amplifier circuit when the base current is $60\,\mu A$ and the collector-emitter voltage of 6 V.

5. An n–p–n transistor passes a collector current of 0.98 mA when the base current in 0.02 mA. Calculate (a) the emitter current and (b) common-emitter current gain.

6. If a common base current gain value is 0.92 determine the common-emitter current gain.

7. An n–p–n transistor has the following characteristics which may be assumed to be linear between the values of the collector voltage stated.

Base current (μA)	20	40	60
Collector current (mA) for $V_{ce} = 2\,V$	1	2	3
Collector current (mA) for $V_{ce} = 4\,V$	1.5	2.6	3.7

The n–p–n transistor is used as a common-emitter amplifier with a load resistor of $2\,k\Omega$ and a collector supply of 8 V. Plot three graphs of collector current (vertically) against collector-emitter voltage. From the graphs determine the common-emitter current gain.

9.11 SMALL SIGNAL AUDIO AMPLIFIER

Aim

To investigate the gain and bandwidth of a small signal amplifier.

Components and equipment

n–p–n transistor, four resistors, three capacitors, signal generator, dual-beam CRO, breadboard and power unit.

Circuit diagram

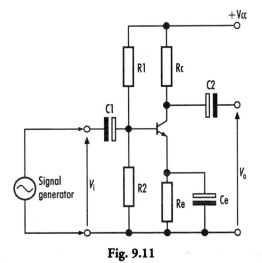

Fig. 9.11

Procedure

(a) Inspect the circuit diagram (Fig. 9.11) and then draw out a wiring diagram on a breadboard sheet.

(b) Get your teacher to check the wiring diagram and then connect up the circuit.

(c) Switch on. Measure the input voltage and output voltage, using the CRO to check that the amplifier circuit is amplifying.

(d) Connect the signal generator and adjust to a frequency of 500 Hz. Check with the CRO that the output waveform is undistorted. Measure the input voltage and the output voltage.

(e) Repeat procedure (d) in steps of 500 Hz until the output waveform distorts.

Table of results

Frequency
Input voltage
Output voltage
Gain

For each set of results calculate gain and plot a graph of gain (vertically) against frequency. Hint: you may need to use log-linear graph paper.

After a discussion with your class teacher and an inspection of your results answer the following questions.

1. State the maximum value of gain obtained.

2. Comment upon the results you have obtained for gain and explain what gain means.

3. Comment upon the shape of the graph you have drawn.

4. Explain why it is necessary to use log-linear graph paper to plot the frequency response curve.

5. Explain what is meant by the term bandwidth.

6. From the graph determine; (a) the lower cut-off frequency; (b) upper cut-off frequency; (c) bandwidth.

7. Give an alternative name for the lower and upper cut-off frequency points.

8. State why it is sometimes necessary to measure gain in bels and decibels.

9. An amplifier has a voltage input of 200 mV and a voltage output of 8 V. Calculate the voltage gain ratio and the gain in decibels.

10. An amplifier has a current input of $2\,\mu\text{A}$ and a current output of 250 mA. Calculate the current gain ratio and the gain in decibels.

11. An amplifier has a power input of 2 mW and an output of 2 W. Calculate the power gain ratio and the power gain in decibels.

9.12 BASIC RELAY CIRCUIT

Aim

To investigate the use of a relay in the control of output devices.

Components and equipment

Relay, diode, signal lamp, motor, bell, buzzer, counter, power unit, breadboard and sufficient batteries to match the needs of the output devices.

Circuit diagram

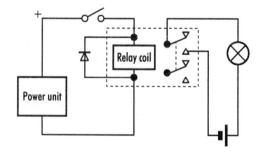

Fig. 9.12

Procedure

(a) Inspect the circuit diagram (Fig. 9.12) and then draw out a wiring diagram on a breadboard sheet.

(b) Get your teacher to check the wiring diagram and then connect up the circuit.

(c) Switch on. Note the effect on the signal lamp. Switch off.

(d) Replace the signal lamp with the motor.

(e) Check whether or not the output device battery needs changing to a more suitable one. If so, change it. Repeat procedure (c).

(f) Repeat procedures (d) and (e) for the other output devices.

After a discussion with your class teacher and an inspection of your results answer the following questions.

1. State what happened to each of the output devices when the switch was closed.

2. Explain the reason for using a relay in such circuits.

3. Do the input voltage and the relay voltage have to be the same value? Give a reason for your answer.

4. State the purpose of the diode in the circuit.

9.13 RELAY LATCHED ALARM CIRCUIT

Aim
To investigate the use of a relay which stays on once triggered.

Components and equipment
Relay, diode, bell, power unit, reset push switch, breadboard.

Circuit diagram

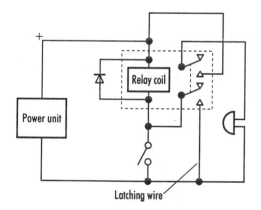

Latching wire

Fig. 9.13

Procedure

(a) Inspect the circuit diagram (Fig. 9.13) and then draw out a wiring diagram on a breadboard sheet.

(b) Get your teacher to check the wiring diagram and then connect up the circuit, but do not connect up the latching wire.

(c) Switch on. Note the effect on the bell. Switch off.

(d) Now connect up the latching wire and note the effect on the bell.

(e) Switch off and again note the effect on the bell.

(f) Now disconnect the latching wire and note the effect on the bell.

(g) Reconnect up the circuit but this time include a reset push switch in series with the latching wire. Switch on and note the effect on the bell. Open the push switch and note the effect on the bell.

After a discussion with your class teacher and an inspection of your results answer the following questions.

1. State the sequence of events that happened during your investigation.

2. State at least one practical use for such a circuit.

3. Design a circuit that could be used for a house alarm to protect one door and one window.

9.14 THE USE OF OUTPUT DEVICES WITH A RELAY

Aim
To investigate the use of transistors in conjunction with a relay to run various output devices.

Components and equipment
Relay, diode, LDR, two *n–p–n* transistors, four resistors, VR, bell, counter, motor, power unit, breadboard and sufficient batteries to meet the needs of the output devices.

Circuit diagram
Fig. 9.14.

Procedure
(a) Inspect the circuit diagram (Fig. 9.14) and then draw out a wiring diagram on a breadboard sheet.
(b) Get your teacher to check the wiring diagram and then connect up the circuit.
(c) Adjust the VR and cover the LDR so that something happens to the bell. Note this effect. Switch off.
(d) Replace the bell with each device in turn and again note what happens.

After a discussion with your class teacher and an inspection of your results answer the following questions.

1. State the sequence of events that happened during your investigation.

2. State the purpose of using transistors in this circuit.

3. State the effect on the resistance of an LDR when it is covered and thereby placed in the dark.

4. Draw a circuit diagram that will allow the LDR to be used as a light-activated switch.

5. Explain why a diode is needed in the circuit.

6. Explain the principle of operation of the electromagnetic counter.

7. What is the purpose of resistor R?

8. What is the purpose of the VR?

9. Draw a block diagram of the system from the details given in the circuit diagram.

10. Design a system that will allow a count to take place of the number of people entering a supermarket entrance doorway.

11. Explain how an LDR can be used in a circuit as an intruder alarm that is used to protect valuable items in a display cabinet.

12. Explain how the circuit can be adapted to be used as a low-temperature sensing device. Include in your answer a block diagram.

13. Explain how the circuit can be adapted to be used as a fire alarm. Include in your answer a block diagram.

14. Instead of just having a fire alarm system installed in a shop what other device could be very usefully activated if a system sensed a fire?

15. Explain, with the aid of a circuit diagram, how an LDR can be used as an intruder alarm and continue to light a signal lamp or continue ringing if it is a bell after the intruder has passed by the LDR.

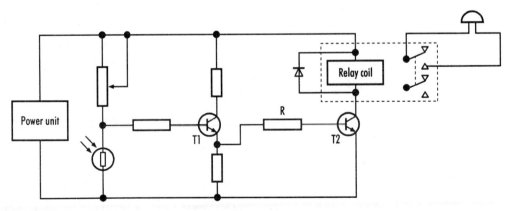

Fig. 9.14

9.15 RADIO RECEIVER

Aim
To build and test a simple radio receiver.

Components and equipment
Two metre length of copper enamelled wire wrapped round a 40 mm diameter cardboard former to act as a coil, 8 m length of wire to act as an aerial, variable capacitor, capacitor, single-point diode, earpiece or headphones, good earth contact point.

Circuit diagram

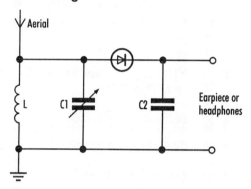

Fig. 9.15

Procedure
(a) Inspect the circuit (Fig. 9.15) and then draw out a wiring diagram on a breadboard sheet.
(b) Get your teacher to check the wiring diagram and then connect up the circuit.
(c) Place the earpiece or headphones in position and then vary the capacitance of the variable capacitor until you can hear a sound.
(d) Vary the capacitance again to check whether other radio stations are available to you.

After a discussion with your class teacher and an inspection of your results answer the following questions.

1. What was the quality of sound like from your first radio station?

2. Could you identify the name of your first radio station?

3. State how many, and the names of, other radio stations received.

4. Explain how the circuit works.

9.16 TUNED RADIO FREQUENCY RECEIVER

Aim
To build and test a tuned radio frequency receiver that operates on a specific frequency.

Components and equipment
Eight metre length of wire to act as an aerial, variable capacitor, three capacitors, single-point diode, inductor, two $n–p–n$ transistors, seven resistors, loudspeaker, breadboard, power unit.

Circuit diagram
Fig. 9.16.

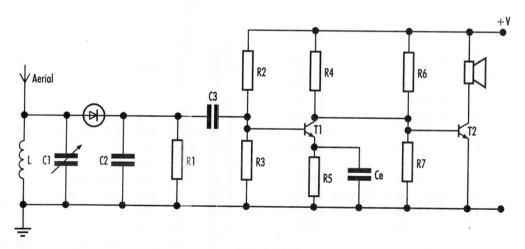

Fig. 9.16

Procedure

(a) In your local newspaper find the section that deals with radio programmes and decide which one you want to tune in to. Write down the frequency value quoted.

(b) Choose a variable capacitor of range 0–300 pF. Calculate the value of the inductor, for a mid-point capacitor value of 150 pF, using $L = 1/(4\pi^2 f^2 C_1)$.

(c) Inspect the circuit diagram (Fig. 9.16) and then draw out a wiring diagram on a breadboard sheet.

(d) Get your teacher to check the wiring diagram and then connect up the circuit.

(e) Switch on and vary the capacitor until you receive a clear signal from the loudspeaker.

After a discussion with your class teacher answer the following questions.

1. Write down the name of the radio station selected. State its frequency band.

2. Write down your workings for the calculated value of the inductor.

3. State the final quality of signal received after varying the capacitor.

4. Draw a block diagram from the information given in the circuit diagram. State the function of each block.

5. State any problems that your encountered in connecting up and making this circuit work. Explain how you overcame these problems.

6. By looking in trade catalogues find a price for each of the components used in your circuit and then calculate the total cost of the radio. Include in your response the amount of time spent on the investigation.

7. A radio receiver tuned acceptor circuit has a variable capacitor and an inductance of 1 H. Determine the setting of the capacitor that will select frequencies of: (a) 6 MHz; (b) 14 MHz.

8. A variable capacitor in a tuned circuit is set to 50 pF. If the value of the inductor is 1 μH determine the frequency of the circuit.

9. A 0.1 μH inductor is connected in parallel with a capacitor to form a circuit that will reject a frequency of 2 GHz but still give a satisfactory band of frequencies. Calculate the value of the required capacitor.

10. A radio receiver tuned acceptor circuit has an inductance of 2 mH and a capacitor which is variable between 200 pF and 1400 pF. Determine the frequency range of this circuit.

9.17 INTRODUCTION TO THE 741 OP AMP

Aim
To investigate the effect of connecting the inverting and non-inverting terminal to 0 V.

Components and equipment
741 op amp, balanced d.c. supply, voltmeter and a breadboard.

Circuit diagram

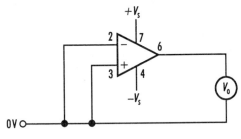

Fig. 9.17

Procedure

(a) Inspect the circuit diagram (Fig. 9.17) and then draw out a wiring diagram on a breadboard sheet.

(b) Get your teacher to check the wiring diagram and then connect up the circuit.

(c) Switch on. Note the output voltage (V_o). Switch off.

(d) Disconnect the inverting terminal from 0 V. Switch on. Note the value of V_o. Switch off.

(e) Reconnect the inverting terminal to 0 V. Disconnect the non-inverting terminal from 0 V. Switch on. Note the value of V_o. Switch off.

After a discussion with your class teacher and an inspection of your results answer the following questions.

1. State the value of V_o when the inverting and non-inverting terminals were connected to 0 V.

2. State the value of V_o when the non-inverting terminal was connected to 0 V.

3. State the value of V_o when the inverting terminal was connected to 0 V.

4. Comment upon the results obtained.

9.18 741 OP AMP OFF-SET VOLTAGE

Aim
To investigate the use of the off-set voltage pins 1 and 5.

Components and equipment
741 op amp, balanced d.c. supply, VR, voltmeter and a breadboard.

Circuit diagram

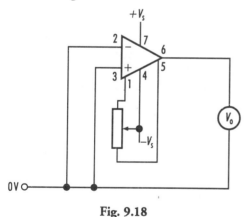

Fig. 9.18

Procedure
(a) Inspect the circuit diagram (Fig. 9.18) and then draw out a wiring diagram on a breadboard sheet.

(b) Get your teacher to check the wiring diagram and then connect up the circuit.

(c) Switch on. Adjust the VR to its minimum setting. Note the output voltage (V_o). Switch off.

(d) Adjust the VR in about five steps to its maximum setting. At each step note the value of V_o.

After a discussion with your class teacher and an inspection of your results answer the following questions.

1. State the value of V_o for each of the variable resistor settings.

2. Comment upon the results obtained.

9.19 CHARACTERISTIC GRAPH FOR A NON-INVERTING OP AMP

Aim
To draw a graph of output voltage (V_o) (vertically) against input voltage (V_i).

Components and equipment
741 op amp, balanced d.c. supply, VR, two resistors, two voltmeters and a breadboard.

Circuit diagram

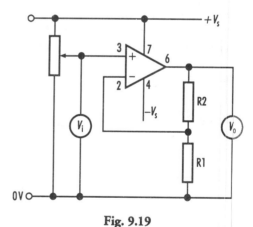

Fig. 9.19

Procedure
(a) Inspect the circuit diagram (Fig. 9.19) and then draw out a wiring diagram on a breadboard sheet.

(b) Get your teacher to check the wiring diagram and then connect up the circuit.

(c) Switch on. Adjust the VR to give V_i and $V_o = 0$. Hint: if you do not get zero on both voltages you may need to use an off-set VR as connected in the previous investigation.

(d) Adjust the VR in about eight steps to give higher input voltages. At each step note the values of V_i and V_o.

(e) Construct a table and record your results.

(f) Repeat procedures (c), (d) and (e) by increasing V_i in the reverse direction. Hint: you may need to reverse the terminals of one of the voltmeters.

(g) Plot a graph of V_i against V_o.

After a discussion with your class teacher and an inspection of your results answer the following question.

1. Comment upon the shape of the graph you have drawn.

9.20 CHARACTERISTIC GRAPH FOR AN INVERTING OP AMP

Aim
To draw a graph of output voltage (V_o) (vertically) against (V_i).

Components and equipment
741 op amp, balanced d.c. supply, VR, two resistors, two voltmeters and a breadboard.

Circuit diagram

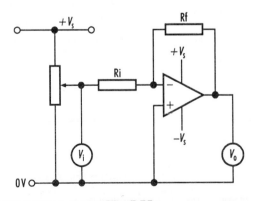

Fig. 9.20

Procedure
(a) Inspect the circuit diagram (Fig. 9.20) and then draw out a wiring diagram on a breadboard sheet.

(b) Get your teacher to check the wiring diagram and then connect up the circuit.

(c) Switch on. Adjust the VR to give V_i and $V_o = 0$. Hint: if you do not get zero on both voltages you may need to use an off-set VR as connected in the previous investigation.

(d) Adjust the VR in about eight steps to give higher input voltages. At each step note the values of V_i and V_o.

(e) Construct a table and record your results.

(f) Repeat procedures (c), (d) and (e) by increasing V_i in the reverse direction. Hint: you may need to reverse the terminals of one of the voltmeters.

(g) Plot a graph of V_i against V_o.

After a discussion with your class teacher and an inspection of your results answer the following question.

1. Comment upon the shape of the graph you have drawn.

9.21 THE VOLTAGE GAIN OF A NON-INVERTING OP AMP

Aim
To investigate and determine the voltage gain of a non-inverting op amp.

Components and equipment
741 op amp, balanced d.c. supply, VR, two resistors, two voltmeters and a breadboard.

Circuit diagram
Fig. 9.21.

Procedure
(a) Inspect the circuit diagram (Fig. 9.21) and then draw out a wiring diagram on a breadboard sheet.

(b) Get your teacher to check the wiring diagram and then connect up the circuit.

(c) Switch on. Adjust the VR to give $V_i = 1$ V. Note the value of V_o. Switch off.

(d) Calculate the gain using Gain $= V_o/V_i$.

(e) Also calculate the gain using Gain $= 1 + (R_2/R_1)$.

(f) Change the feedback resistor R2 to one of a higher value.

(g) Repeat procedures (c), (d), (e) and (f).

(h) Again change the value of the feedback resistor R2 to one of a higher value and repeat the procedures again. Do this for six values of the feedback resistor.

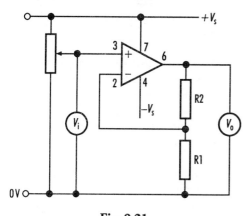

Fig. 9.21

Table of results

	Gain 1	Gain 2	Gain 3	Gain 4	Gain 5	Gain 6
V_i						
V_o						
R1						
R2						

After a discussion with your class teacher and an inspection of your results answer the following questions.

1. Make a comment about the gain obtained using the voltages as compared to the gain obtained using the resistors.

2. What conclusion do you reach from such results?

3. The op amp shown in Fig. 9.22 works on a $\pm 9\,\text{V}$ supply. (a) Is it an inverting or non-inverting amplifier? (b) Calculate the voltage gain. (c) If an a.c. of peak value $+200\,\text{mV}$ is applied to the input, calculate the peak output voltage.

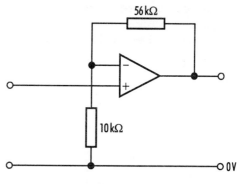

Fig. 9.22

4. The circuit in Fig. 9.23 shows an op-amp operating a bell. (a) Calculate the voltage at A. (b) Calculate the voltage at B when V_i is (i) $+6\,\text{V}$, (ii) $+3\,\text{V}$, (iii) $0\,\text{V}$, (iv) $-3\,\text{V}$. (c) State the range of voltage that will operate the bell.

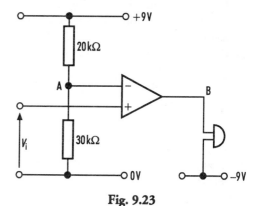

Fig. 9.23

5. The power supply to a non-inverting op amp is $\pm 15\,\text{V}$ with $V_i = +2\,\text{V}$. The feedback resistor is $30\,\text{k}\Omega$ and the other resistor is $10\,\text{k}\Omega$. Calculate the value of V_o.

6. A non-inverting op amp has $V_i = 2\,\text{V}$ with a power supply of $\pm 15\,\text{V}$. The feedback resistor is $300\,\text{k}\Omega$ and the other resistor is $10\,\text{k}\Omega$. Calculate the value of V_o.

9.22 THE VOLTAGE GAIN OF AN INVERTING OP AMP

Aim

To investigate and determine the voltage gain of an inverting op amp.

Components and equipment

741 op amp, balanced d.c. supply, VR, two resistors, two voltmeters and a breadboard.

Circuit diagram

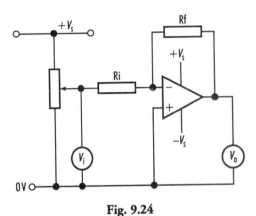

Fig. 9.24

Procedure

(a) Inspect the circuit diagram (Fig. 9.24) and then draw out a wiring diagram on a breadboard sheet.

(b) Get your teacher to check the wiring diagram and then connect up the circuit.

(c) Switch on. Adjust the VR to give $V_i = 1\,\text{V}$. Note the value of V_o. Switch off.

(d) Calculate the gain using Gain $= -V_o/V_i$.

(e) Also calculate the gain using Gain $= -R_f/R_i$.

(f) Change the feedback resistor Rf to one of a higher value.

(g) Repeat procedures (c), (d), (e) and (f).

(h) Again change the value of the feedback resistor Rf to one of a higher value and repeat the procedure again. Do this for six values of the feedback resistor.

Table of results

	Gain 1	Gain 2	Gain 3	Gain 4	Gain 5	Gain 6
V_i						
V_o						
R_i						
R_f						

After a discussion with your class teacher and an inspection of your results answer the following questions.

1. Make a comment about the gain obtained using the voltages as compared to the gain using the resistors.

2. What conclusion do you reach from such results?

3. Explain what is meant by the term inverting op amp.

4. Calculate the gain in an inverting op amp if the feedback resistor is $160\,\text{k}\Omega$ and the input resistor is $10\,\text{k}\Omega$.

5. The circuit diagram and characteristic graph for a voltage amplifier are shown in Fig. 9.25. (a) State which part of the graph is in its linear stage. (b) Calculate the voltage gain for the linear region. (c) Calculate the value of the input resistor if the feedback resistor is $18\,\text{k}\Omega$. (d) Find from the characteristic graph the value of the peak a.c. output voltage if the a.c. input peak voltage is $1\,\text{V}$.

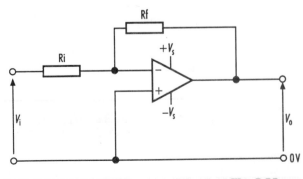
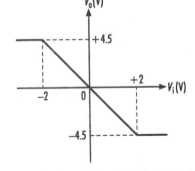

Fig. 9.25

6. The input waveform to the voltage amplifier in Fig. 9.25 has a peak value of 1 V. Draw the input waveform and the output waveform if the gain for the amplifier is 3.

7. The voltage amplifier being used in Fig. 9.25 is working on a ± 5 V supply. The input a.c. waveform has a peak value of 1.5 V. Draw the input waveform and the output waveform for a gain of 4.

8. Draw a circuit diagram to show how an op amp can be connected to give a gain of (a) -2, (b) -1, (c) $+21$, (d) $+5$. In each case use a value of $2 \mathrm{k\Omega}$ or less for the lower value resistor.

9.23 LIGHT-SENSITIVE OR TEMPERATURE-SENSITIVE CIRCUITS USING AN OP AMP

Aim
To investigate the use of an LDR and a thermistor to control an output lamp using an op amp.

Components and equipment
741 op amp, d.c. supply, LDR, thermistor, n–p–n transistor, VR, four resistors, signal lamp and a breadboard.

Circuit diagram
Fig. 9.26.

Procedure
(a) Inspect the circuit diagram (Fig. 9.26) and then draw out a wiring diagram on a breadboard sheet.
(b) Get your teacher to check the wiring diagram and then connect up the circuit.
(c) Switch on. Adjust the VR so that the signal lamp is just on. Place your hand over the LDR and note what happens. Switch off.
(d) Interchange the LDR with the VR. Switch on. Does the signal lamp light? Now place your hand over the LDR. Does the signal lamp light? Switch off.
(e) Remove the LDR and replace the VR in its original position. Connect the thermistor in the original position of the LDR.
(f) Switch on. Adjust the VR so that the signal lamp just goes off. Gently heat the thermistor and note what happens to the signal lamp. Switch off.

After a discussion with your class teacher and an inspection of your results answer the following questions.

1. For procedure (c) what happens to the signal lamp when the LDR is in the dark?

2. What do the initials NTC mean when related to a thermistor?

3. What do the initials PTC mean when related to a thermistor?

4. State the effect on an LDR of light falling upon it.

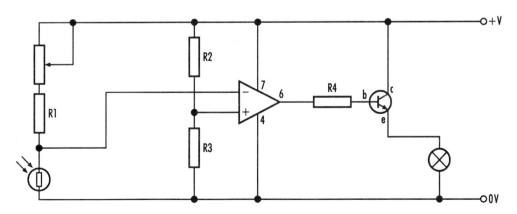

Fig. 9.26

5. State the effect of a thermistor being heated.

6. For procedure (d) state what happened when the LDR was interchanged with the VR.

7. For procedure (f) state what happened to the signal lamp when the thermistor was gently heated.

8. Complete the following tables:

LDR	Signal lamp	Thermistor	Signal lamp
Dark		Cold	
Light		Hot	

9. State at least one application for the use of an LDR and a thermistor in this manner.

10. An op amp is being used in conjunction with a thermistor to act as a 'cold' alarm in a greenhouse. The circuit diagram and a characteristic graph as shown in Fig. 9.27. The signal lamp should light when the temperature just falls below 0°C. (a) State an alternative signalling device to the signal lamp that will emit sound. (b) Explain why a sound device is often better than a display device. (c) From the graph determine the temperature when the resistance is 24 Ω. (d) From the graph determine the resistance of the thermistor when the temperature is (i) + 6°C, (ii) 0°C, (iii) −6°C. (e) Calculate the voltage at A when the thermistor is at 0°C. (f) State the voltage that the variable resistor should be set at for the signal lamp to operate correctly. (g) State what happens to the voltage at A when the thermistor is cooled from + 6°C to − 6°C. (h) State what happens to the signal lamp when the temperature is dropping as stated in part (f).

9.24 THE SCHMITT TRIGGER SWITCHING CIRCUIT USING AN OP AMP

Aim
To investigate the time taken for switching actions.

Components and equipment
741 op amp, d.c. supply, VR, three resistors, LED, voltmeter with wandering lead and a breadboard.

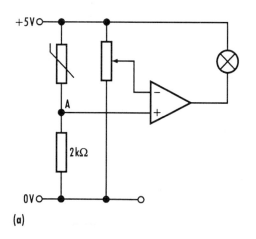

(a)

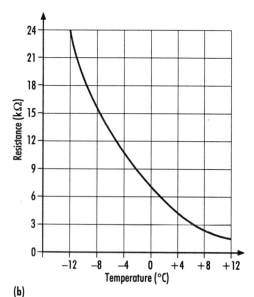

(b)

Fig. 9.27

Circuit diagram

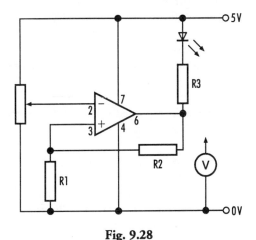

Fig. 9.28

Procedure

(a) Inspect the circuit diagram (Fig. 9.28) and then draw out a wiring diagram on a breadboard sheet.

(b) Get your teacher to check the wiring diagram and then connect up the circuit.

(c) Switch on. Adjust the VR to its minimum position. Note the state of the LED.

(d) Check with the wandering voltmeter lead that the output voltage is very near to the supply voltage.

(e) Measure the voltage at pin 3 and confirm that it is approximately 50% of the supply voltage.

(f) Adjust the VR until the LED lights up. Measure the voltage at pins 2, 3 and 6.

After a discussion with your class teacher and an inspection of your results answer the following questions.

1. Write a description of what happened to your circuit when each step of the procedure was carried out.

2. Is the Schmitt trigger circuit an astable or bistable circuit? Give reasons for your answer.

3. Draw the characteristic of a Schmitt trigger which has an upper trip point (UTP) of 4 V, a lower trip point (LTP) of 2 V, a high state (HS) of 15 V and a low state (LS) of 5 V.

4. A 6 V a.c. input is applied to the Schmitt trigger in the previous question. Draw the output voltage waveform.

5. State at least one other application for a Schmitt trigger circuit.

9.25 MONOSTABLE TIMER 555

Aim
To investigate the use of a one stable state timer.

Components and equipment
555 timer, d.c. supply, push switch, electrolytic capacitor, three resistors, LED, buzzer and a breadboard.

Circuit diagram

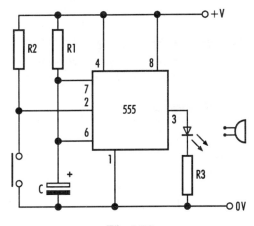

Fig. 9.29

Procedure

(a) Inspect the circuit diagram (Fig. 9.29) and then draw out a wiring diagram on a breadboard sheet.

(b) Get your teacher to check the wiring diagram and then connect up the circuit.

(c) Switch on and note what happens to the LED. Hint: wait for a fairly long time otherwise you will miss the point of this investigation. Switch off.

(d) Replace the LED with a buzzer and repeat procedure (c).

After a discussion with your class teacher and an inspection of your results answer the following questions.

1. State what happened to the LED and the buzzer after the switch was closed.

2. Explain how the circuit works.

3. Calculate the time that an LED stays on in a circuit as shown in Fig. 9.29 if resistor $R1 = 2\,M\Omega$ and capacitor $C = 5\,\mu F$. Use time $(T) = 1.1R_1 C$.

4. How many states does a monostable have?

5. State at least one practical application for such a circuit.

9.26 ASTABLE TIMER 555

Aim
To investigate the use of a system that has no stable state.

Components and equipment
555 timer, d.c. supply, switch, electrolytic capacitor, three resistors, LED, and a breadboard.

Circuit diagram

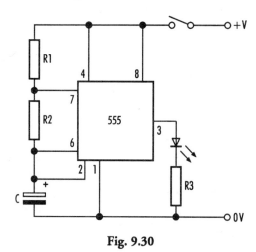

Fig. 9.30

Procedure
(a) Inspect the circuit diagram (Fig. 9.30) and then draw out a wiring diagram on a breadboard sheet.
(b) Get your teacher to check the wiring diagram and then connect up the circuit.
(c) Switch on and note what happens to the LED. Hint: wait for a fairly long time

otherwise you will miss the point of this investigation.
(d) Switch off.

After a discussion with your class teacher and an inspection of your results answer the following questions.

1. State what happened to the LED after the switch was closed.

2. Explain how the circuit works.

3. Calculate the frequency of oscillation for the circuit if resistor $R1 = 2\,k\Omega$, $R2 = 62\,k\Omega$ and $C = 10\,\mu F$.

4. State two other names for the 555 timer being used in this way.

5. State at least one practical application for such a circuit.

6. How many states does an astable have?

9.27 THE 555 ELECTRONIC ORGAN

Aim
To build and test an electronic organ.

Components and equipment
555 timer, d.c. supply, eight preset VRs, one VR, two resistors, three capacitors, diode, switch earpiece or loudspeaker and a breadboard.

Circuit diagram
Fig. 9.31.

Information
The 555 timer is being used as an astable oscillator. A square wave is produced at pin 3. The frequency of oscillation depends upon the capacitance C_1 and the total resistance between pins 6 and 7 and between pins 7 and 8. The preset VRs allow a range of frequencies to be obtained and thus allow each note to be tuned to a separate frequency. When the metal probe is placed on any of the preset VRs it will be connected to pins 7 and 8. For different preset connections there will be a change in the

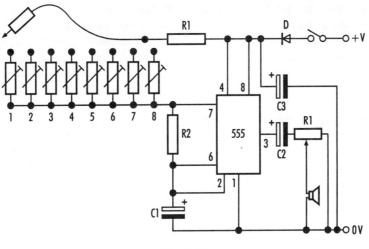

Fig. 9.31

frequency of oscillation. The preset VRs give us our keyboard.

Procedure
(a) Inspect the circuit diagram (Fig. 9.31) and then draw out a wiring diagram on a breadboard sheet.
(b) Get your teacher to check the wiring diagram and then connect up the circuit.
(c) Switch on. Connect the metal probe to each preset VR in turn. A sound, i.e. a note, should be heard from positions 1 to 8 inclusive.
(d) Adjust each preset VR until the note required is achieved. A musical instrument or tuning fork should be used to carry out this process.
(c) Use the VR next to the loudspeaker to adjust the volume.

9.28 LOGIC GATES – THE HEX INVERTER OR NOT GATE

Aim
To investigate the characteristics of a NOT gate.

Components and equipment
NOT gate, d.c. supply, LED, resistor, switch and breadboard.

Circuit diagram

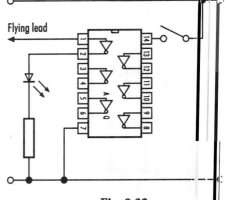

Flying lead

Fig. 9.32

Procedure
(a) Inspect the circuit diagram (Fig. 9.32) then draw out a wiring diagram on a breadboard sheet.
(b) Get your teacher to check the wiring diagram and then connect up the circuit.
(c) Connect the flying lead to the +V. Switch on and note the effect on the Switch off. Remove the flying lead.
(d) Connect the flying lead to the 0V. Switch on and note the effect on the Switch off. Remove the flying lead.

After a discussion with your class teacher inspection of your results answer the following questions.

1. Write down the catalogue number of the NOT gate used in this investigation.

2. The symbol used in the circuit diagram was American. Draw a BS graphical symbol for a NOT gate.

3. State what happened to the LED during the investigation.

4. Complete the truth table, using 1 when the LED came on and 0 if the LED did not come on.

Input	Output Q
High (1) + V	
Low (0) 0 V	

Comment upon the results obtained.

9.29 LOGIC GATES – THE AND GATE

Aim
To investigate the characteristics of an AND gate.

Components and equipment
AND gate, d.c. supply, LED, resistor, switch and breadboard.

Circuit diagram

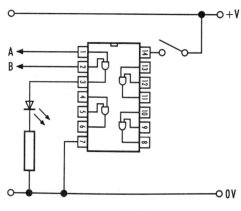

Fig. 9.33

Procedure
(a) Inspect the circuit diagram (Fig. 9.33) and then draw out a wiring diagram on a bread-board sheet.

(b) Get your teacher to check the wiring diagram and then connect up the circuit.

(c) Connect the flying leads A and B to the 0 V rail. Switch on and note the effect on the LED. Switch off. Remove the flying leads.

(d) Connect the flying leads A to the 0 V rail and B to the 5 V rail. Switch on and note the effect on the LED. Switch off. Remove the flying leads.

(e) Connect the flying leads A to the 5 V rail and B to the 0 V rail. Switch on and note the effect on the LED. Switch off. Remove the flying leads.

(f) Connect the flying leads A and B to the 5 V rail. Switch on and note the effect on the LED. Switch off. Remove the flying leads.

After a discussion with your class teacher and an inspection of your results answer the following questions.

1. Write down the catalogue number of the AND gate used in this investigation.

2. The symbol used in the circuit diagram was American. Draw a BS graphical symbol for an AND gate.

3. State what happened to the LED during the investigation.

4. Complete the truth table, using 1 when the LED came on and 0 if the LED did not come on.

Input		Output Q
A	B	
0	0	
0	1	
1	0	
1	1	

5. Comment upon the results obtained.

6. State at least one practical application for logic gates.

9.30 LOGIC GATES – THE NAND GATE

Aim
To investigate the characteristics of a NAND gate.

Components and equipment
NAND gate, d.c. supply, LED, resistor, switch and breadboard.

Circuit diagram

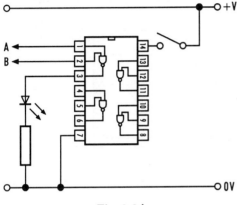

Fig. 9.34

Procedure
(a) Inspect the circuit diagram (Fig. 9.34) and then draw out a wiring diagram on a breadboard sheet.
(b) Get your teacher to check the wiring diagram and then connect up the circuit.
(c) Connect the flying leads A and B to the 0 V rail. Switch on and note the effect on the LED. Switch off. Remove the flying leads.
(d) Connect the flying leads A to the 0 V rail and B to the 5 V rail. Switch on and note the effect on the LED. Switch off. Remove the flying leads.
(e) Connect the flying leads A to the 5 V rail and B to the 0 V rail. Switch on and note the effect on the LED. Switch off. Remove the flying leads.
(f) Connect the flying leads A and B to the 5 V rail. Switch on and note the effect on the LED. Switch off. Remove the flying leads.

After a discussion with your class teacher and an inspection of your results answer the following questions.

1. Write down the catalogue number of the NAND gate used in this investigation.

2. The symbol used in the circuit diagram was American. Draw a BS graphical symbol for a NAND gate.

3. State what happened to the LED during the investigation.

4. Complete the truth table, using 1 when the LED came on and 0 if the LED did not come on.

Input		
A	B	Output Q
0	0	
0	1	
1	0	
1	1	

5. Comment upon the results obtained.

6. State at least one practical application for logic gates.

9.31 LOGIC GATES – THE OR GATE

Aim
To investigate the characteristics of an OR gate.

Components and equipment
OR gate, d.c. supply, LED, resistor, switch and breadboard.

Circuit diagram

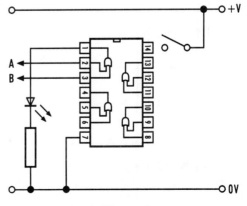

Fig. 9.35

Procedure

(a) Inspect the circuit diagram (Fig. 9.35) and then draw out a wiring diagram on a breadboard sheet.

(b) Get your teacher to check the wiring diagram and then connect up the circuit.

(c) Connect the flying leads A and B to the 0 V rail. Switch on and note the effect on the LED. Switch off. Remove the flying leads.

(d) Connect the flying leads A to the 0 V rail and B to the 5 V rail. Switch on and note the effect on the LED. Switch off. Remove the flying leads.

(e) Connect the flying leads A to the 5 V rail and B to the 0 V rail. Switch on and note the effect on the LED. Switch off. Remove the flying leads.

(f) Connect the flying leads A and B to the 5 V rail. Switch on and note the effect on the LED. Switch off. Remove the flying leads.

After a discussion with your class teacher and an inspection of your results answer the following questions.

1. Write down the catalogue number of the OR gate used in this investigation.

2. The symbol used in the circuit diagram was American. Draw a BS graphical symbol for an OR gate.

3. State what happened to the LED during the investigation.

4. Complete the truth table, using 1 when the LED came on and 0 if the LED did not come on.

Input		
A	B	Output Q
0	0	
0	1	
1	0	
1	1	

5. Comment upon the results obtained.

6. State at least one practical application for logic gates.

9.32 LOGIC GATES – THE NOR GATE

Aim
To investigate the characteristics of a NOR gate.

Components and equipment
NOR gate, d.c. supply, LED, resistor, switch and breadboard.

Circuit diagram

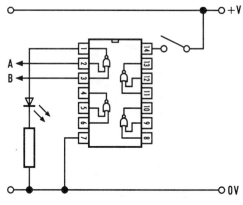

Fig. 9.36

Procedure

(a) Inspect the circuit diagram (Fig. 9.36) and then draw out a wiring diagram on a breadboard sheet.

(b) Get your teacher to check the wiring diagram and then connect up the circuit.

(c) Connect the flying leads A and B to the 0 V rail. Switch on and note the effect on the LED. Switch off. Remove the flying leads.

(d) Connect the flying leads A to the 0 V rail and B to the 5 V rail. Switch on and note the effect on the LED. Switch off. Remove the flying leads.

(e) Connect the flying leads A to the 5 V rail and B to the 0 V rail. Switch on and note the effect on the LED. Switch off. Remove the flying leads.

(f) Connect the flying leads A and B to the 5 V rail. Switch on and note the effect on the LED. Switch off. Remove the flying leads.

After a discussion with your class teacher and an inspection of your results answer the following questions.

1. Write down the catalogue number of the NOR gate used in this investigation.

2. The symbol used in the circuit diagram was American. Draw a BS graphical symbol for a NOR gate.

3. State what happened to the LED during the investigation.

4. Complete the truth table, using 1 when the LED came on and 0 if the LED did not come on.

Input		
A	B	Output Q
0	0	
0	1	
1	0	
1	1	

5. Comment upon the results obtained.

6. State at least one practical application for logic gates.

9.33 LOGIC GATES – THE QUAD 2-INPUT POSITIVE NAND GATE

Aim
To investigate the use of a NAND gate to make other gates.

Components and equipment
Quad 2-input positive NAND gate, d.c. supply, LED, resistor, switch and breadboard.

Circuit diagram
Fig. 9.37.

Procedure
(a) Inspect the circuit diagram (Fig. 9.37(a)) and then draw out a wiring diagram on a breadboard sheet.
(b) Get your teacher to check the wiring diagram and then connect up the circuit.
(c) Connect the flying leads A and B to the 0 V rail. Switch on and note the effect on the LED. Switch off. Remove the flying leads.

(d) Connect the flying leads A to the 0 V rail and B to the 5 V rail. Switch on and note the effect on the LED. Switch off. Remove the flying leads.
(e) Connect the flying leads A to the 5 V rail and B to the 0 V rail. Switch on and note the effect on the LED. Switch off. Remove the flying leads.
(f) Connect the flying leads A and B to the 5 V rail. Switch on and note the effect on the LED. Switch off. Remove the flying leads.
(g) Repeat procedures (a) to (f) for the NAND gates shown in Figs 9.37(b) and 9.37(c).

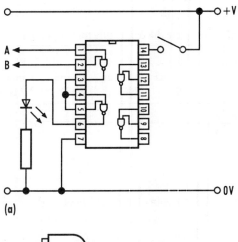

(a)

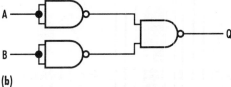

(b)

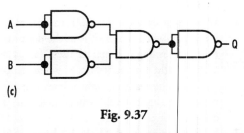

(c)

Fig. 9.37

After a discussion with your class teacher and an inspection of your results answer the following questions.

1. Write down the catalogue number of the Quad 2-input positive NAND gate used in this investigation.

2. For each investigation state what happened to the LED.

3. For each investigation, complete a truth table, using 1 when the LED came on and 0 if the LED did not come on.

Input		
A	B	Output Q
0	0	
0	1	
1	0	
1	1	

4. Complete the truth table for the gates listed.

Input		Output OR	Output NOR	Output AND	Output NAND
A	B				
0	0				
0	1				
1	0				
1	1				

5. Look at your results and look at the truth table and state for each investigation your findings.

9.34 SUGGESTED ANSWERS FOR INVESTIGATIONS

Section 9.2

1. When light shines on the LDR its resistance is low and when the LDR is in the dark the resistance is high. Practical uses: (a) as a parking light for roadworks – when darkness falls the warning light comes on; (b) measuring light levels when used in a camera.
2. LDR resistance in the light is about $1\,k\Omega$ and in the dark about $10\,M\Omega$.
3. Before the signal lamp will light it now has to be darker.
4. The signal lamp comes on more slowly in the dark because some of the original current is being used to charge the capacitor.

5. If the capacitor is incorrectly connected, it will not work and could itself be damaged and also damage components in the circuit.
6. The signal lamp now comes on in the light and goes off in the dark. Practical uses: (a) as an intruder detection system; (b) as part of a counting system in an industrial conveyor belt situation.
7. The level of light could be varied according to the position of the slider on the variable resistor.
8. Using two transistors instead of one makes the circuit more sensitive to smaller changes in light level. The two transistors are called a Darlington pair amplifier.
9. In light the LDR resistance is low. Current from the positive of the battery takes the easiest route back to the negative of the battery which is through the LDR and not through the transistors. In dark the LDR resistance is high. Current from the positive of the battery cannot go through the LDR so it goes through the base of transistor 1, through transistor 2. The base currents 'switch on' the transistors and if the collector currents of both transistors together are large enough, then the signal lamp lights.

Section 9.3

1. The signal lamp came on.
2. The resistance of the thermistor decreases when it is heated. The VR resistance is high to the thermistor resistance. The current flowing from the positive of the battery must then go through the first transistor into the base of the second transistor. The current is amplified and if large enough will light the signal lamp.
3. Fire alarm, etc.
4. The signal lamp came on.
5. The resistance of the thermistor increases when it is cooled. The VR resistance is low relative to the thermistor resistance. The current flowing from the positive of the battery must then go through the first transistor into the base of the second transistor. The current is amplified and if large enough will light the signal lamp.
6. Frost alarm etc.

7. See figure.

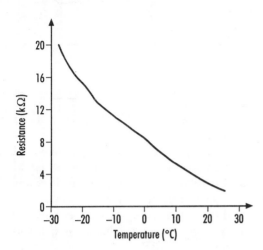

(a) $2.8\,k\Omega$, $4.8\,k\Omega$, $8\,k\Omega$, $10.8\,k\Omega$, $16\,k\Omega$. (b) $9°C$, $-8°C$, $18.5°C$. (c) At $25°C$, power $= 25 \times 0.4 = 10\,mW$; at $8°C$, power $= 8 \times 0.4 = 3.2\,mW$. (d) An NTC thermistor is so manufactured that with increase in temperature the thermistor resistance decreases. Likewise with a decrease in temperature the resistance increases.

8. Positive temperature coefficient (PTC) thermistor. With this thermistor the resistance increases as the temperature increases.

9. $P = V^2/R = 6^2/2200 = 0.01\,636$ W $=$ $16.36\,mW$. The power dissipated in the bead does not exceed the $500\,mW$ maximum.

10. $P = V^2/R = 35^2/2200 = 0.5568$ W $=$ $556.8\,mW$. The power dissipated in the bead does exceed the $500\,mW$ maximum.

11. The thermistor is likely to be damaged and other circuit components could also be damaged.

Section 9.4

1. The lamp does not light.
2. The lamp will light signalling that the liquid is at a maximum level and any further increase in depth could be dangerous.
3. When there is a conducting liquid between the probes, the circuit is complete. Resistor R is high relative to the resistance of the liquid between the probes. The current

flowing from the positive of the battery must then go through the first transistor into the base of the second transistor. The current is amplified and if large enough will light the signal lamp.

4. Applications: to detect whether or not it is raining outside, liquid level indicators in any size of container from a glass for a partially sighted person to a large vat in the brewing industry, body resistance measure, i.e. it measures the amount of perspiration on a person's hands, etc.

Section 9.5

1. The signal lamp will flash on and off.
2. The signal lamp flashes on and off because each transistor, in turn, is switched on and off, due to capacitor C1 charging and discharging through resistor R2, and capacitor C2 charging and discharging through resistor R3.
3. During procedure (e) using a smaller capacitor in position C1 the signal lamp will still flash on and off but the lamp is off for longer than it is on.
4. The flashing rate depends upon the values of $C_1 R_2$ and $C_2 R_3$. In this case $C_1 R_2$ would be a smaller numerical value than $C_2 R_3$.
5. During procedure (h) both lamps flashed on and off. The lamps would flash one after the other as each transistor is switched on and off in turn.
6. The circuit is called an astable multivibrator or a free-running multivibrator. It has no stable states but switches from a high state to a low state automatically and vice versa at a rate determined by the circuit component values. It generates square wave pulses and is called a relaxation oscillator.
7. Because the lamps flash on and off it is very useful as a warning device for roadworks, school crossings, navigation lamps at sea and similar applications.
8. If resistor R2 is changed to a smaller value then the on time is longer than the off time in the original circuit.
9. $f = 1/(1.4 \times 0.2 \times 10^{-6} \times 10 \times 10^3) =$ $357\,Hz$.

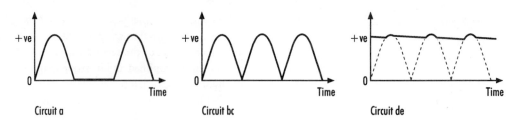

Circuit a   Circuit bc   Circuit de

Section 9.6

1. A sound came from the loudspeaker.
2. No sound came from the loudspeaker.
3. In light, the resistance of the LDR is low, so the transistors switch on and off which produces a sound in the loudspeaker. In the dark, the LDR resistance is high, so the transistors do not switch on and off. Therefore, there is no sound. The circuit is called an astable multivibrator. A typical application is a burglar alarm. The device can be placed in a dark room and anyone entering with light will set off the loudspeaker.
4. The sound from the loudspeaker would be lower in tone.

Section 9.7

1. During procedures (c) and (d) the signal lamp would come on after a time delay. Different settings give different delay times.
2. As the capacitor values increased there would be a longer time delay at each step.
3. Practical applications include any situation where a specific amount of time is needed for a particular process, e.g. egg timer, photographic timer, etc.

Section 9.8

1. The drawing should be the shape of a sine wave.
2. The purpose of the fuse is for protection and the switch is used to control the a.c. supply.
3. The transformer is used to step-down the supply voltage from 230 V a.c. to a value that will not destroy the diode circuits.
4. In normal power supplies the output voltage is often 12 V.
5. $N_2 = N_1 V_2/V_1 = (1200 \times 20)/230 = 104.3$, say 105.
6. See figure (top).
7. Circuit Fig. 9.7(a) – one diode gives half-wave rectification. Circuit Fig. 9.7(b) – two diodes give full-wave rectification but an expensive centre-tapped transformer is needed. Circuit Fig. 9.7(c) – four diodes give full-wave rectification using a transformer which costs less than a centre-tapped one. Circuit Fig. 9.7(d) – by using a capacitor the waveforms are smoothed out. Circuit Fig. 9.7(e) – by using a zener diode the output voltage remains constant and will not vary despite fluctuations in the supply voltage.
8. It is easier to understand the waveform from circuit diagram Fig. 9.7(c) if it is divided into two as shown in the figure (bottom).

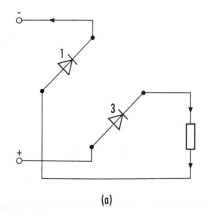

(a)

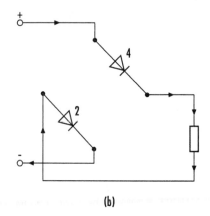

(b)

The diodes are so arranged in the bridge circuit so that in the circuit in Fig. (a) the current flows through diodes 1 and 3 and in the circuit in Fig. (b) the current flows through diodes 2 and 4. The current flow, in total, through the load resistor is unidirectional.

9. When a zener diode is reverse biased to its breakdown zener voltage, the voltage across it stays constant for a wide range of reverse currents. Because of this property, a zener diode can be used to stabilize the output voltage from a smoothed circuit.

10. (a) Zener diode current $= (9-4)/100 = 0.05$ A or 50 mA. (b) (i) Zener diode current $= (12-4)/100 = 0.08$ A or 80 mA. (ii) Zener diode current $= (3-4)/100 =$ a negative answer, so the current is zero, i.e. the supply voltage must always be higher than the zener breakdown voltage.

11. (a) Maximum zener current $= P/V = 0.5/5.2 = 0.09615$ A. (b) Potential difference across the resistor $= 11.5 - 5 = 6.5$ V. (c) Resistance $= 6.5/0.09615 = 67.6\,\Omega$. From E12 series resistance would be 68 Ω.

Section 9.9

1. See Fig. 7.27.
2. See Fig. 7.27.
3. See 7.10 Example 3.
4. See 7.10 Example 3.
5. See Table 7.1.
6. When the input current is zero, the output current should be zero, but in practice there is a very small leakage current which is due to minority carriers. In modern silicon transistors this is extremely low.

Section 9.10

1. As the base current increases so does the collector current. The current gain should be approximately the same for each calculation per graph.
2. Write down the quoted value, e.g. AC127 – 50 at a collector current of 500 mA, BC108 125 at a collector current of 2 mA, N3706 – 600 at a collector current of 0 mA.
3. All transistors of the same number vary slightly because of tolerance in manufacture.

4. Current amplification factor $= I_c/I_b$ $(2 \times 10^{-3})/60 \times 10^{-6}) = 33.3$.
5. Emitter current = collector current + base current $= 0.98 + 0.02 = 1$ mA. Common emiter current gain $= I_c/I_b = 1/0.02 =$
6. Common-emitter current gain $= 0.$ $(1 - 0.92) = 11.5$.
7. On the graph of collector current against collector-emitter voltage, a load line ne be plotted to determine a value for V_c an The load line equation is $V_{cc} = V_{ce} + I$ When $I_c = 0$ $V_{cc} = V_{ce} = 8$ V. X $V_{ce} = 0$ then $I_c = V_{cc}/R_L = 8/20($ 0.004 A $= 4$ mA. The load line is 1 from $V_{cc} = 8$V to $I_c = 4$ mA. Points A B are marked on the graph where the line cuts the graphs at base currents of 2 and 60 μA. The collector current is sured as 1.3 mA with a base current of 4(i.e. the difference between 60 and 2(Common-emitter current gain $I_c/I_b = (1.3 \times 10^{-3})/(40 \times 10^{-6}) = 32.5$.

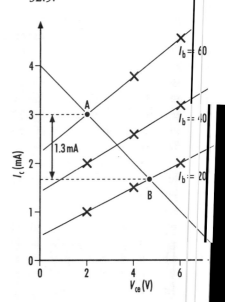

Section 9.11

1. Accept any sensible value for gain.
2. The gain increases, then levels out starts to decrease. When defining performance a sine wave signal i considered. The amplifier is cons be operating in a linear manner Th

gain describes the characteristic of an amplifier and is equal to V_2/V_1.

3. The expected shape of the graph is shown:

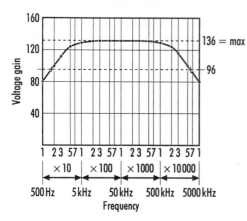

4. Log-linear graph paper is used so that a very wide range of frequency can be plotted. Look at the scale used in the graph in question 3.

5. Amplifiers are placed in a class by stating a range of signal frequencies over which they are designed to provide a specific gain. This specified range of frequencies is known as the bandwidth.

6. Calculate the position of 0.707 of the voltage gain as shown on the graph. Gain at 0.707 maximum = $0.707 \times 136 = 96$. In this case the lower cut-off frequency is about 800 Hz with an upper cut-off frequency of about 3000 kHz. Bandwidth = upper cut-off frequency – lower cut-off frequency = $3\,000\,000 - 800 = 2\,999\,200$ Hz.

7. 3 dB points.

8. Sometimes the numbers involved in voltage gain become high and difficult to deal with. The range of frequency can also be very wide and become difficult to plot on graph paper. By using a logarithmic scale for gain the problems are overcome. Voltage gain in decibels = $20 \log_{10}(V_2/V_1)$.

9. $V_1 = 200\,\text{mV} = 0.2\,\text{V}$, $V_2 = 8\,\text{V}$.
Voltage gain ratio = $V_2/V_1 = 8/0.2 = 40$.
Voltage gain (dB) = $20 \log 40 = 20 \times 1.60\,206 = 32.04$.

10. $I_1 = 2\,\mu\text{A} = 0.000\,002\,\text{A}$; $I_2 = 250\,\text{mA} = 0.25\,\text{A}$. Current gain ratio = $I_2/I_1 = 0.25/0.000\,002 = 125\,000$. Current gain (dB) = $20 \log 125\,000 = 20 \times 5.0969 = 101.94$.

11. $P_1 = 2\,\text{mW} = 0.002\,\text{W}$, $P_2 = 2\,\text{W}$. Power gain ratio = $P_2/P_1 = 2/0.002 = 1000$. Power gain (dB) = $10 \log 1000 = 10 \times 3 = 30$. (Watch this calculation because for power gain in decibels we use $10 \log(P_2/P_1)$ – ask your teacher to explain this.)

Section 9.12

1. When the switch was closed the output devices, i.e. signal lamp, motor, bell, buzzer and counter, all worked.

2. A relay can be considered to be a remote-controlled switch. It is a switch which is turned on and off by an electromagnet. A relay allows an input supply to operate the relay and switch on an output device using a different supply.

3. The relay output voltage can be different from the input because they are isolated from one another. There are no conducting wires linking the input with the output.

4. A diode is always included in relay circuits because it protects other components. When the current in the relay drops to zero, a large voltage is induced in the coil due to its inductance. This voltage, known as a back e.m.f., could damage any components used to control the current in the coil. The most likely component to get damaged is a transistor. In this circuit a transistor has not been used but it is good practice to get into the habit of always using a diode for protection. The diode is connected in reverse bias so that it gives an easy route for the induced e.m.f. thus avoiding any damage to the circuit.

Section 9.13

1. Procedure (c) – bell rings but stops ringing when switched off. Procedures (d), (e) and (f) – bell continues to ring whether or not the switch is on after the bell has started to ring; the circuit is now 'latched' and will continue to ring until the power unit is switched off or the latch wire removed. Procedure (g) – the push switch allows the circuit to be reset, i.e. the bell can be switched off.

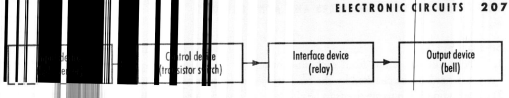

| Input device | Control device (transistor switch) | Interface device (relay) | Output device (bell) |

and moves it on by one position. Each movement represents one digit.

7. Resistor R is there to prevent excessive base currents in transistor T2 destroying it when transistor T1 is off and the voltage at the collector of transistor T1 is approaching the supply voltage of the circuit.

8. The VR allows the light level at which switching occurs to be changed.

9. See figure (top).

10. On the left-hand side of the doorway place the counter circuit of Fig. 9.14 and on the right-hand side of the doorway place a signal lamp that is illuminated at all times the count is needed and can be 'seen' by the LDR. When a customer walks through the entrance the beam of light is broken and the counter operates.

11. On the left-hand side of the cabinet place the bell circuit of Fig. 9.14 and on the right-hand side of the cabinet place a signal lamp that is illuminated at all times and can be 'seen' by the LDR. When the light source is broken by a hand being placed in the cabinet the bell will ring.

12. Use the circuit diagram in Fig. 9.14 but use a thermistor instead of an LDR. See figure (bottom).

13. Interchange the thermistor with the VR. The block diagram is the same as that in question 12.

14. A sprinkler system. The design could include automatically turning on a solenoid valve that controls water sprinklers.

15. Use the same circuit as Fig. 9.14 but include a latching circuit as used in Fig. 9.13. Try it, but before switching on get your teacher to check the circuit. It is the second set of relay contacts that allows the current to by-pass the second transistor and hold the relay on even after the intruder has moved away from the LDR.

| Input device | Control device | Interface device (relay) | Warning device (bell/buzzer) |

Section 9.15

1. The quality of sound from each student's radio needs to be checked by the teacher.
2. The radio station name needs to be checked by the teacher.
3. The number of stations and their names need to be checked by the teacher.
4. The tuning circuit is the inductor L and the variable capacitor C1. The tuning circuit will respond very strongly to a.c. signals of one particular frequency which is known as the resonance frequency. This resonance frequency depends upon the values of the coil and the variable capacitor.

   The detecting circuit is the diode and capacitor C2. The diode rectifies the amplitude-modulated radio-frequency current received from the tuning circuit. This rectification then leaves the positive half-cycles of the current which contain the useful audio frequencies and the useless radio frequencies. The useless radio frequencies are filtered by capacitor C2 to earth and the useful audio frequencies go on to the earpiece or headphones and can be heard. By further variation of capacitor C1 the resonant frequency can be changed to tune in to other radio stations.

   In the circuit it is important to use a good aerial and a very sound earth connection and a very sensitive earpiece or headphone.

Section 9.16

1. The teacher should check the details listed for each radio station.
2. The teacher should check the individual workings for the inductance value.
3. Again the teacher should check the quality of the final signal received by each radio in the class.
4. See figure (bottom). The aerial intercepts the radio wave at a particular frequency. The tuning circuit is the inductor and the variable capacitor, which respond to a.c. signals of one frequency. The demodulator

is the diode and capacitor C2. The diode rectifies the amplitude-modulated radio-frequency current received from the tuning circuit. As in the previous investigation, we are only interested in the audio frequencies that go on for further treatment. The demodulator introduces minimum distortion but provides an adequate output. The small signal amplifier does exactly as its name suggests, the signal and sends it on its way, undistorted, into the buffer amplifier. The buffer amplifier is an amplifier stage designed to isolate a preceding circuit from the loading effects of a following circuit. A buffer stage provides an impedance transformation between the two circuits. The last part of the circuit is a loudspeaker, often in the form of an earpiece or a set of headphones.

6. The figure quoted for the cost of the tuned radio receiver by each student will have to be checked by the teacher for each student in the class.
7. (a) Consider a frequency of 6 MHz then: $C = 1/(4\pi^2 f^2 L) = 1/(4\pi^2)(6 \times 10^6)^2 \times (1 \times 10^{-6}) = 703.4\,\text{pF}$. (b) Likewise when $f = 14\,\text{MHz}$ then $C = 129.2\,\text{pF}$.
8. $f = 1/[2\pi\sqrt{(LC)}] = 1/2\pi\sqrt{(1 \times 10^{-6} \times 0.1 \times 50 \times 10^{-12})} = 22.5\,\text{MHz}$.
9. $C = 0.0633\,\text{pF}$.
10. Frequency range is 7.957 MHz to 3.007 MHz.

Section 9.17

1, 2 and 3 The output voltage (V_o) will be the positive or negative supply voltage. When one of the input terminals is removed there might be no change in V_o or there could be a complete reversal in the sign of V_o.

4. The difference between the inverting terminal voltage and the non-inverting terminal voltage will be zero. V_o for this situation should be zero. But usually there is a small off-set voltage on one of the input terminals which will produce a difference of slightly

larger than zero, the sign of which is unknown. This difference voltage is amplified causing V_o to be saturated.

Section 9.18
1. The output voltage (V_o) in all cases is zero.
2. The op amp has a large open-loop gain so the smallest value of off-set voltage will lead to saturation in either the positive or negative direction.

Section 9.19
1. The shape of the graph can be found in Fig. 7.40. V_o is directly proportional to the V_i over a range of input voltages if the graph is linear. The graph then shows that after a particular range of input voltages are exceeded saturation takes place as indicated by the horizontal lines.

Section 9.20
1. The shape of the graph obtained is a mirror image of the one drawn in the previous investigation, i.e. it has been inverted, hence the name of this type of circuit.

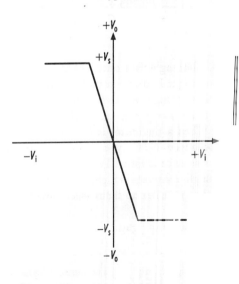

Section 9.21
1. The calculated values of gain using voltages and the resistors should in case by very close.

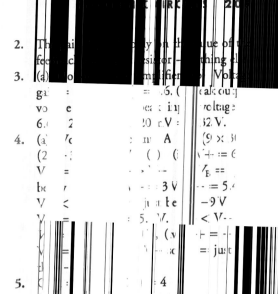

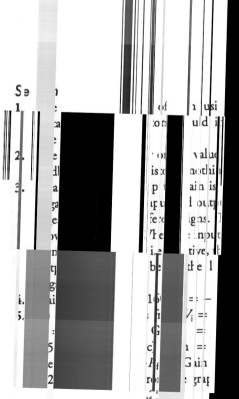

6.

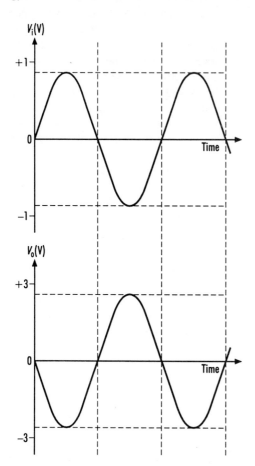

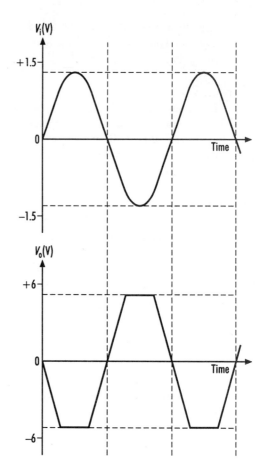

7. $V_o = 4 \times 1.5 = 6\,\text{V}$. The input is large enough to saturate the op amp. The output is then distorted with 'flattened' peaks. (figure top right).

8. (a) Inverting op amp. See Fig. 9.24. Use $R_f = 4\,\text{k}\Omega$ and $R_i = 2\,\text{k}\Omega$. (b) Inverting amplifier. See Fig. 9.24. Use $R_f = 2\,\text{k}\Omega$ and $R_i = 2\,\text{k}\Omega$. (c) Non-inverting op amp. See Fig. 9.21. Use $R_f = 40\,\text{k}\Omega$ and $R_i = 2\,\text{k}\Omega$. (d) Non-inverting amplifier. See Fig. 9.21. Use $R_f = 8\,\text{k}\Omega$ and $R_i = 2\,\text{k}\Omega$.

Section 9.23

1. The signal lamp goes out.

2. Negative temperature coefficient – this type of thermistor has a low resistance when hot.

3. Positive temperature coefficient – this type of thermistor has a high resistance when hot.

4. When light falls on an LDR its resistance changes. In complete darkness the resistance is very high but in full sunlight the resistance is very low, e.g. $1\,\text{k}\Omega$.

5. When a thermistor is heated its resistance changes: NTC type – temperature high, resistance is low; PTC type – temperature high, resistance is high.

6. The signal lamp lights when now in darkness.

7. Signal lamp lights.

8.

LDR	Signal lamp	Thermistor	Signal lamp
Dark	off	Cold	off
Light	on	Hot	on

9. The LDR can be used in several different types of alarm systems. The thermistor can be used in any system that needs to give a signal when the temperature is falling below or going above a pre-calculated value.

10. (a) Bell, buzzer or siren. (b) Sound can usually be heard but sometimes light is not so easily seen. (c) When the resistance is $24\,\Omega$ the temperature is about $-12°C$. (d) $R_{+6} = 3\,k\Omega$, $R_o = 7.5\,k\Omega$, $R_{-6} = 3.2\,k\Omega$. (e) $V_A = (5 \times 2)/(7.5 + 2) = 1.0526$ V. (f) 1.05 V. (g) The voltage drops 1.98 V to 0.62 V. (h) Initially the signal lamp is off. As the voltage drops below 1 V the lamp comes on.

Section 9.24

1. When the voltage at pin 2 is low the output voltage at pin 6 is high. The LED is off and the voltage at pin 3 is approximately 50% of the supply value. When the voltage at pin 3 rises, the LED will suddenly come on. At this time the output voltage at pin 6 drops quickly to 0 V.

2. The Schmitt trigger is a very important bistable circuit. It is designed to change state when the input or trigger voltage level goes above a predetermined value and changes back to its original state when the input voltage falls below that predetermined value.

3.

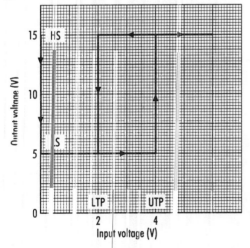

4.

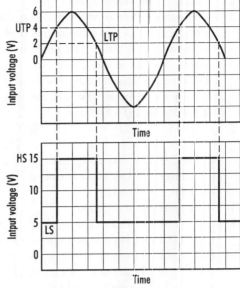

5. The Schmitt trigger can be used as a debouncing switch circuit, any situation which needs level detection and for reshaping pulse-type waveforms.

Section 9.25

1. When the switch is closed the LED lights for a period of time and then goes out. The same sequence of events happens with the buzzer.

2. With the switch open, the output is 0 V, so the LED is off. When the switch is closed, the output is changed to $+V$, say 15 V, the LED is on and alight. The capacitor starts to charge and when the voltage on pins 6 and 7 reach 67% of the supply voltage, the output returns to 0 V, its stable state. The LED goes out.

3. $T = 1.1R_1C = 1.1 \times 2 \times 10^6 \times 5 \times 10^{-6} = 11$ s.

4. A monostable is a one-state timer which always returns to its original state after a period of time. This time lapse depends on the values of resistor R and capacitor C1.

5. A visual or sound alarm.

Section 9.26

1. The LED flashes on and off.
2. The astable circuit is very similar to the monostable except that with the astable pin 2 is connected to pin 6. This produces an automatic restart to the timing cycle. The astable output is never still; it will switch between two voltages, 0 V and the maximum positive supply voltage.
3. $f = 1.44/(R_1 + 2R_2)C = 1.44/(2000 + 2 \times 62\,000)10 \times 10^{-6} = 1.143\,\text{Hz}$. This frequency means that in this case the LED is flashing on and off at the rate of just over one cycle per second.
4. Pulse generator or oscillator.
5. Any situation where flashing lights are needed. One example is as a safety device on a bicycle. A flashing lamp at the rear, in the dark, could more easily be seen than just a single light. Another application is by changing the LED for a loudspeaker. This again could be used as an alarm system.
6. An astable has no stable states. As shown in the investigation the LED switches on and off showing that it is changing state all of the time.

Section 9.28

1. Accept any correct catalogue number.
2. See Fig. 7.44.
3. When the input was high the output was low and vice versa.
4.

Input	Output Q
High (1) + V	0
Low (0) 0 V	1

The gate has a single input and a single output. A high (1) output is produced when the input is low (0). A low output (0) is produced when the input is high (1).

Section 9.29

1. Accept any correct catalogue number.
2. See Fig. 7.44.
3. The LED only came on once.

Input		Output Q
A	B	
0	0	0
0	1	0
1	0	0
1	1	1

5. With an AND gate, if both inputs are low (0) the output is low (0). If only one input is high (1) the output remains low (0). If both inputs are high (1) then the output goes high (1).
6. Accept any correct practical application.

Section 9.30

1. Accept any correct catalogue number.
2. See Fig. 7.44.
3. The LED stayed off only once.
4.

Input		Output Q
A	B	
0	0	1
0	1	1
1	0	1
1	1	0

5. With a NAND gate, if both inputs are low (0) the output is high (1). If only one input is high (1) the output remains high (1). If both inputs are high (1) then the output goes low (0).
6. Accept any correct practical application.

Section 9.31

1. Accept any correct catalogue number.
2. See Fig. 7.44.
3. The LED stayed off only once.
4.

Input		Output Q
A	B	
0	0	0
0	1	1
1	0	1
1	1	1

5. With an OR gate if both inputs are low (0) the output is low (0). If both inputs are high (1) the output is high (1). If one input is high (1) the output is high (1).
6. Accept any correct practical application.

Section 9.32

1. Accept any correct catalogue number.
2. See Fig. 7.44.
3. The LED came on only once.
4.

Input		Output
A	B	Q
0	0	1
0	1	0
1	0	0
1	1	0

5. With a NOR gate if both inputs are low (0) the output is high (1). If both inputs are high (1) the output is low (0). If one input is high (1) the output is low (0).
6. Accept any correct practical application.

Section 9.33

1. Accept any correct catalogue number.
2. Fig. 9.37 – the LED came on only when both inputs were high. Fig. 9.37(b) – the LED was on at all times except when both inputs were low. Fig. 9.37(c) – the LED came on only when both inputs were low.

3.

Input		Output Fig. 9.37(a)	Output Fig. 9.37(b)	Output Fig. 9.37(c)
A	B			
0	0	0	0	1
0	1	0	1	0
1	0	0	1	0
1	1	1	1	0

4.

Input		Output OR	Output NOR	Output AND	Output NAND
A	B				
0	0	0	1	0	1
0	1	1	0	0	1
1	0	1	0	0	1
1	1	1	0	1	0

5. The circuit diagram in Fig. 9.37(a) produced an AND gate. The circuit diagram in Fig. 9.37(b) produced an OR gate. The circuit diagram in Fig. 9.37(c) produced a NOR gate.

INDEX

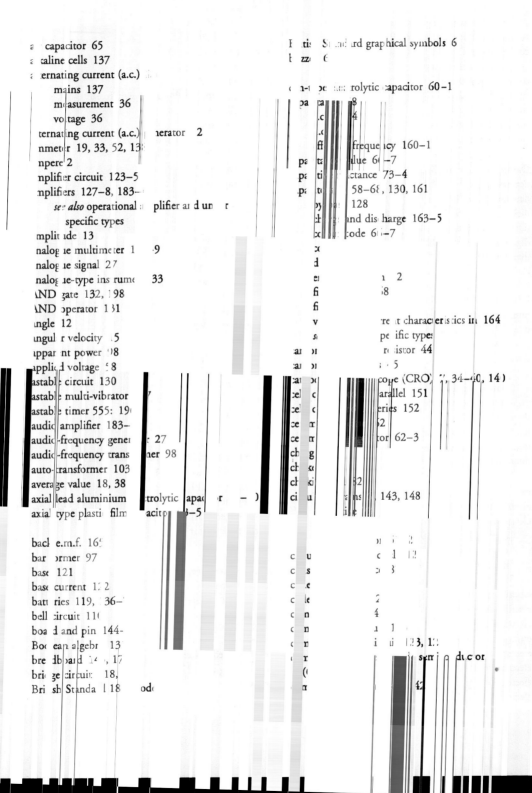

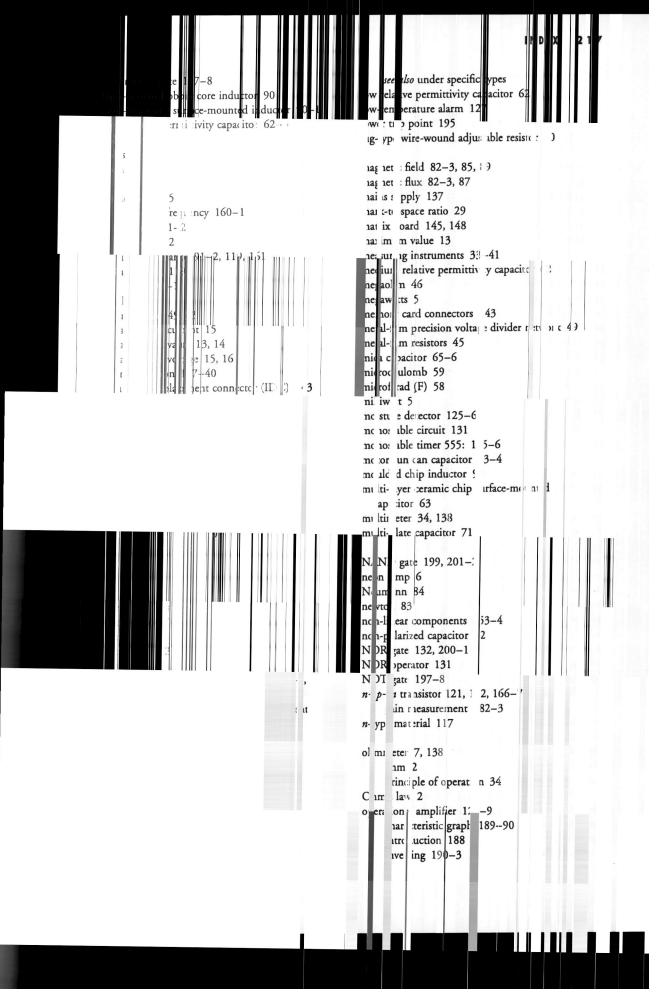